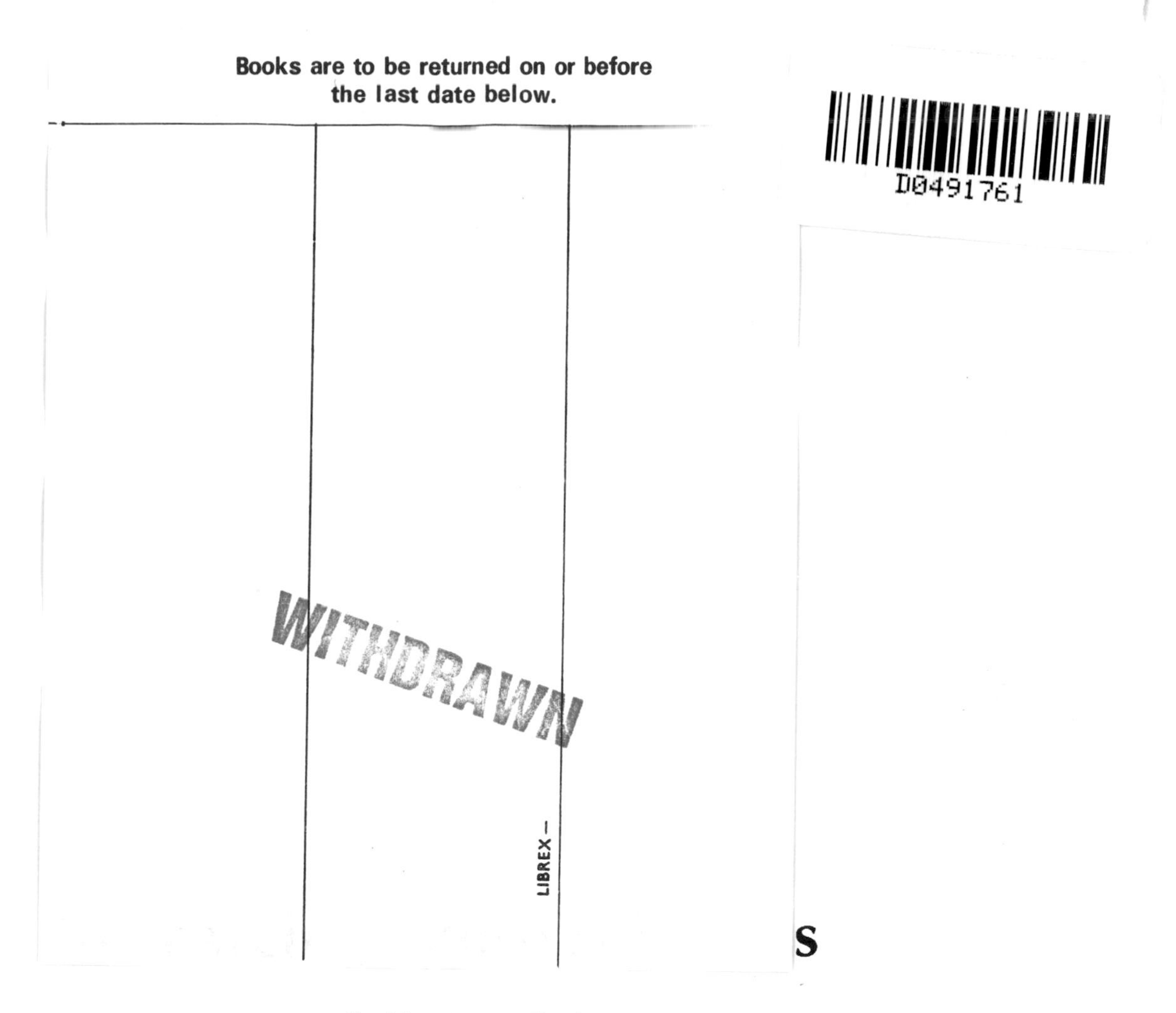

S

Compiled by Janet Taylor

Teacher Fellow
The Royal Society of Chemistry
1992–93

In Search of more Solutions

Compiled by Janet Taylor

Edited by Catherine O'Driscoll, Helen Eccles and Neville Reed

Designed by Imogen Bertin

Published by the Education Division, The Royal Society of Chemistry

Printed by The Royal Society of Chemistry

For further information on other educational activities undertaken by the Royal Society of Chemistry write to:

The Education Department
The Royal Society of Chemistry
Burlington House
Piccadilly
London W1V OBN

ISBN 1 870343 35 2

British Library Cataloguing in Publication Data.
A catalogue for this book is available from the British Library.

Foreword

This collection of problem–solving activities for students on post-16 courses is another example of curriculum material written by teachers of chemistry for teachers of chemistry. The project follows on from the Society's first collection of egg races and other problem solving activities in chemistry. *In search of solutions* was sent to every secondary school in January 1991.

As President of the Royal Society of Chemistry, I would very much like thank all the people who sent in ideas, and the willing volunteers who tested the activities in schools and colleges. I think that the level of enthusiasm and innovation shown by teachers and lecturers bodes well for the future of chemistry.

Professor Charles Rees CChem FRSC FRS
President, The Royal Society of Chemistry
July 1994

Esso

Contents

Introduction

Janet Taylor

Now there is one outstandingly important fact regarding Spaceship Earth, and that is that no instruction book came with it. ...So we are forced, to use our intellect, which is our supreme faculty, to devise scientific experimental procedures and to interpret effectively the significance of experimental findings.

R. Buckminster Fuller
Operating manual for spaceship earth, 1969

The inspiration behind the Royal Society of Chemistry's *In search of solutions* was a concern that chemistry was missing out in the competitions[1] known as egg races that flourished in the early 1980s. Great egg races in the lower school and technological competitions in the sixth form were giving students that 'little bit extra' to motivate them to follow a scientific or technological career.[2] Most of the egg-races involved the application of the principles of physics and engineering. The quest was on to bring the excitement of competition to school chemistry but it soon became clear that chemical egg races were hard to invent.

The collection of activities in the first book grew to include enjoyable non-competitive problem-solving activities, all aiming to put the fun back into chemistry. This book builds on the first publication and develops and extends the philosophy to meet the needs of students who are, in the main, following a post-16 chemistry course. The activities have been trialled in schools and colleges around the UK.

It became clear, as the book developed, that student-centred learning activities are in demand. They have become an integral part of many post-16 courses. I hope that this book will be a useful resource for teachers looking for ideas for assignments and investigations.

The book is a collection of activities that provide students with opportunities to develop problem-solving skills. The Royal Society of Chemistry has recently published a book that addresses the teaching of problem solving in chemistry.[3] My remit has been to produce a bank of resources for chemical problem solving. The activities, which are mainly practical, are presented as tasks. The students can design their own experiments and it is hoped that as they overcome the obstacles within the tasks they will become more inventive and enthusiastic.

I have tried to include 'real' chemical problems, some of which may not fit in with current perceptions of problem solving in schools and colleges. There is a point in all problem-solving chains and cycles where ideas need to be generated and the students have to rely on what is already known. In 'real' chemistry, even the post-16 target audience is unlikely to have sufficient background knowledge. Sometimes a practical method will need to be suggested. This information is usually included in the teacher's section of the activity. It is not my intention to suggest that there are unique solutions to these problems but rather to provide hard-pressed teachers with at least one tried and tested approach. References are included.

Some of the activities have been used as student-centred learning assignments. Evaluation or assessment can sit comfortably within a problem-solving approach.[4] The tasks may be broken down, either by the teacher or the students, into a series of goals corresponding to the objectives of the course. Then, in the following evaluation or assessment of performance, the results and conclusions can be matched against these goals. An effective problem solver will have attributes that tend to fall outside assessment schemes. Solutions will show creativity, originality, elegance, economy,

Esso

I am greatly indebted to the many people who contributed to this book. I have tried to acknowledge the originator of the problems where they could be identified. Contributions took many forms, ranging from free samples of equipment to consultations with scientists in specialist areas. I would like to thank everyone who helped in any way. I am especially grateful to the members of the Society who offered to test activities at the outset of the project. Their response was very heartening.

I envisage that the reader will dip into this book to find an idea or possibly a solution. All of the activities stand alone. The activities have been categorised under various headings in the contents pages. The activities are ordered: those requiring post-16 chemical knowledge are at the front and towards the back are problems that require some lateral thinking! I hope that some of them will fulfil the original aim of putting the fun back into chemistry.

1. K. Davies, *In search of solutions.* London: RSC, 1990.

2. B. E. Woolnough, *The making of engineers and scientists.* Oxford: Oxford University Department of Educational Studies, 1991.

3. C. Wood, *Creative problem solving in chemistry.* London: RSC, 1993.

4. *Problem solving with industry.* Sheffield: Sheffield Hallam University, 1991.

Acknowledgements

Janet Taylor

I would like to thank the following people for their help, inspiration and time.

Bob Aveyard	School of Chemistry, University of Hull
John Barker	Centre for Educational Studies, King's College, London
Eric Bates	Comino Foundation
Peter Borrows	Waltham Forest
Joe Burns	Coleg Trydyddol Glan Hafren Tertiary College, Wales
Justin Dillon	Centre for Educational Studies, King's College, London
Bill Harrison	Centre for Science Education, Sheffield Hallam University
Andrew Hunt	Nuffield Science in Practice project, London
Colin Johnson	The Chemiquest Project, Techniquest, Cardiff
Alex Johnstone	Centre for Science Education, University of Glasgow
Robert Hadden	Centre for Science Education, University of Glasgow
Roger Locke	School of Education, University of Birmingham
Dean Madden	National Centre for Biotechnology Education, University of Reading
John Scholar	National Centre for Biotechnology Education, University of Reading
John Sellars	Formerly Business and Technology Education Council
Dave Sant	Centre for Science Education, Sheffield Hallam University
Rod Wardlaw	Sheffield Hallam University
Mike Watts	Roehampton Institute

The help of the technical staff, and especially Andrew Walden, at NESCOT, Epsom's College of Further and Higher Education has been greatly appreciated.

Dr Neville Reed, Education Officer (Schools and Colleges), directed the project and I would like to thank him for his guidance and enthusiasm. I would also like to thank all of the staff in the Society's Education Department for their kindness and help during my year as Teacher Fellow.

I would like to join the Royal Society of Chemistry in thanking the staff and students of the following schools and colleges for their help in trialling the experiments in this book:

The American Community School, Cobham, Surrey
The Arnewood School, New Milton, Hampshire
Christ the King Sixth Form College, Lewisham
Christleton High School, Christleton, Chester
The City Technology College, Kingshurst, Birmingham
Coleg Glan Hafren, Rumney, Cardiff
Cranbrook School, Cranbrook, Kent
Eltham College, London
Gloucestershire College of Arts and Technology, Cheltenham, Gloucestershire
Gorseinon College, Gorseinon, Swansea
Greenhead College, Huddersfield
Hammersmith and West London College, London
Hove Park School, Hove, East Sussex
Ilford County High School, Barkingside, Essex

Leek High School, Leek, Staffordshire
Loughborough Grammar School, Loughborough, Leicestershire
Loughborough High School, Loughborough, Leicestershire
Mill Hill School, Mill Hill, London
NESCOT, Epsom's College of Further and Higher Education, Surrey
Porth County Comprehensive School, Porth, Mid Glamorgan
Richmond upon Thames Tertiary College, Twickenham, Middlesex
Sandwell College, Smethwick, Birmingham
St Aloysius College, Hornsey Lane, London
St Edward's School, Oxford
St George's School for Girls, Edinburgh
St Paul's School, Barnes, London

The Royal Society of Chemistry would also like to thank the many individuals who have given their time and effort to this project, in particular:

David Andrews, Berinda Banks, Alida Burdet, Peter Calder, Bruce Dixon, Hugh Dunlop, Neil Heeley, Susan Jackson, Eric Lewis, Ann Lewis-Kell, Allan Moir, David Moore, Bob Mudd, Alan Neuff, Colin Osborne, Mike Parker, Ian Poots, Desrine Price, Mike Roderick, Philip Rodgers, Neil Tunstall, Martin Wesley and Peter Wright.

The Society would like to acknowledge the following for their help with the photographs used in the book.

The late Professor Charles Giles (p10).
Ann-Marie Colbert and Paul Spurdens (p43).
Paul Spurdens (p180).

The Society wishes to thank Surrey Local Education Authority and the governors of NESCOT for seconding Janet Taylor to the Society's Education Department.

Thanks are also due to Esso UK plc for helping to fund the production of this publication.

Index of activities with brief descriptors

1 **Only dust – is there a sign of life?**
Making a polarimeter to analyse a sample of dust for enantiomeric properties.

2 **Finding the 'rate expression' for the reaction between iodine and tin**
Developing an experimental technique to research a problem in reaction kinetics.

3 **Discover the properties of an element**
Patterns in chemical and physical properties across the Periodic Table. The element is silicon and the problem may be extended to investigate thermal effects on the electrical resistivity of a semiconductor.

4 **The hunt for vitamin C; the effect of cooking processes on the vitamin C content of cabbage**
Devising effective sampling techniques to make a quantitative comparison of vitamin C content.

5 **Nobili's rings**
An electrochemical phenomenon.

6 **From milk to curds and whey – which enzyme?**
An investigation into which of three enzymes is best for making curds and whey from milk.

7 **Any glucose?**
Designing a biosensor to detect glucose.

8 **Degrees of acidity**
The temperature rise when a known amount of a strong acid reacts with a given excess of a strong alkali is used to determine whether the acid is monoprotic or diprotic.

9 **Franklin's teaspoon of oil**
This is a problem with an historical setting. Benjamin Franklin experimented with thin oil films on Clapham Common pond in the 1770s. His interpretation stopped short of a simple calculation that might have led him to speculate on the size of particles of matter. The problem is to complete these calculations and confirm his findings.

10 **Which solution is which?**
Qualitative inorganic analysis on a series of salts.

11 **Are these jelly babies natural?**
The extraction and chromatography of food dyes.

12 **A transient red colour: the aqueous chemistry between iron(III) ions and sulphur oxoanions**
An unexplained phenomenon in inorganic chemistry.

13 **An analysis of coloured sands**
This is an opportunity for some practical geochemistry. Students with a special interest in geology may like to do some preliminary research into the nature and formation of different sands. Red, yellow and brown sands owe their colour to the presence of iron(III) oxide. Several analytical methods are available to determine how much iron is present in samples of coloured sand.

THE ROYAL
SOCIETY OF
CHEMISTRY

Esso

Esso

THE ROYAL
SOCIETY OF
CHEMISTRY

46 **An elementary problem**
A logic problem based on industrial processes.

47 **Making ice**
Investigating the quickest way to make ice.

48 **Fizzy drinks**
Investigating what makes a drink froth.

49 **After meal puzzle**
A practical puzzle which is solved by using the depression of the freezing

50 **A chemically powered boat: a bubble boat race**
Designing and making a model boat that is propelled by bubbles produced by the reaction between sodium hydrogencarbonate and citric acid. In egg race terminology this reaction is known as the 'chemical rubber band'. A race may be held to evaluate the designs.

Index of activities grouped by general contexts

In this index some activities appear under more than one heading.

Unexplained phenomena and open problems

Challenges, competitions and egg-races

Design and make

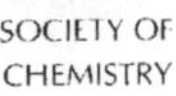

Esso

Problems based on daily life and work

Identifying unknown substances

Detection and analysis

Investigations in physical chemistry

Historical problems

Theoretical problems

Index of activities grouped by topics

In this index some activities appear under more than one heading.

The Periodic Table

Structure and bonding

Electrochemistry

Energy transfer: chemical energy to mechanical energy

Energy transfer: chemical energy to heat energy

Phase changes

Esso

THE ROYAL
SOCIETY OF
CHEMISTRY

Chemistry of gases

Acids and bases

Amino acids

Carbohydrates

Vitamins

Dyes

Polymers

Enzymes

Safety

Peter Borrows

In open-ended problem-solving activities it may be difficult to anticipate all of the strategies that students might adopt in attempting to solve a particular problem. The best solutions are often the completely unexpected ones: the imagination of young people, uncluttered with a knowledge of the 'right' answers, may be much more original than that of their teachers. It is therefore wise to set problems that require only relatively low hazard chemicals or procedures.

Teachers need to be particularly vigilant during practical problem-solving activities. A higher degree of supervision is needed than in activities with more closed outcomes. Students may need to be questioned about what they are doing. This is no bad thing as it will help with the assessment of their performance on a problem. Students need to be encouraged to take a responsible attitude towards safety, both of their own and that of others, and a statement to that effect should appear prominently in the instructions for the problem. In planning their solution to the problem, students should be asked to consider safety. In appropriate cases, they might be asked to carry out their own risk assessment. A possible *pro forma* to help them do this is included on page xix; this may be copied. Even if this is used, however, teachers must still check the plans before they are implemented. Remember, however, that you are not checking whether the plan will work, but whether it is safe. Bear in mind that students often do not stick to their plans: reasonably enough, these are modified in the light of experience. Constant vigilance is therefore necessary, to prevent new hazards from being introduced.

Always insist on eye protection. Even if a student still claims to be thinking, the student on the opposite side of the room may not be! During a competitive event, and especially if artifacts are being tested in some way at the end, excitement levels can rise. Do not allow things to get out of hand, and stop competitors from putting themselves or others in danger.

Under the *COSHH regulations* there is an obligation on employers to carry out a risk assessment whenever chemicals hazardous to health are used or made. The *Management of safety at work regulations* require similar risk assessment whenever any potentially hazardous activity is carried out. You will therefore need to check students' proposed plans against your employer's risk assessments. Most education employers have followed the recommendations of the Health and Safety Commission in *COSHH: guidance for schools*, and have adopted various nationally available publications as the basis for their General Risk Assessments. The most commonly used are:

Hazcards. Uxbridge: CLEAPSS School Science Service, 1989. (A new edition in preparation 1994 and is only available to members.)

Topics in safety, 2nd edn. Hatfield: ASE,1988.

Hazardous chemicals, A manual for schools & colleges. Edinburgh: SSSERC/Oliver & Boyd, 1979.

Safeguards in the school laboratory, 9th edn. Hatfield: ASE, 1988.

Laboratory handbook. Uxbridge: CLEAPSS School Science Service, 1989 and later supplements [only available to members].

Between them these publications cover most of the situations likely to be met in schools and colleges, but not necessarily situations that will be met in problem-solving activities. Where the General Risk Assessments do not suffice, the employer should have set up a procedure for Special Risk Assessments. This is likely to involve subscribers contacting the CLEAPSS School Science Service. Sometimes, a teacher will be able to adapt General Risk Assessments, on the basis of chemical analogy. Particularly useful guidance on carrying out risk assessments in open-ended situations will be found in *Preparing COSHH risk assessments for project work in schools.* Edinburgh: SSERC, 1991.

Before students begin any practical activity, teachers should always check that what the student is proposing to do is compatible with the employer's risk assessments, whether contained in some or all of the above publications or based on the employer's local rules, or the result of a special risk assessment.

Pro forma sheet for risk assessment

RISK ASSESSMENT FORM

Name of person completing this form: ______________________________

Outline of proposed practical procedure: ______________________________

____

____

____

Chemical used or made for potentially hazardous procedure	Hazard (*eg* flammable, corrosive, toxic; exposure limit *etc*)	Source(s) of advice/ general risk assessments (*eg Hazcards, Topics in Safety etc*)	Strategy to reduce risk (*eg* substitute chemicals; reduce scale; use fume cupboard, safety screens, protective gloves, goggles *etc*)

Junk list

Many of the chemical egg races type of task require the use of 'junk'. The following is a list of the types of item envisaged:

plastic lemonade bottles
'squeezy' bottles (washing up liquid containers)
empty beer/soft drink cans (dry)
coffee tins/syrup tins
coffee jars/jam jars
yoghurt pots/margarine tubs
shoe boxes/cereal packets
cardboard tubes from toilet rolls/kitchen towels
blocks of expanded polystyrene packing
polystyrene meat trays/egg boxes
disposable foil trays (oven ready)
lollipop sticks
wood off-cuts/cotton reels
used tights

In chemical egg races non junk items are often used alongside junk:

sticky tape
glue and/or glue gun
Blu-tack/plasticine
string
rubber bands
paper clips/split fasteners
pegs
wire
pins
aluminium kitchen foil
cling film
balloons
plastic bags
drinking straws
plastic tubing
assorted bungs and corks
plastic syringes
plastic gloves
paper towels
stapler
ruler
simple tools: tin snips, saw, bradawl, file, stanley knife *etc.*

THE ROYAL
SOCIETY OF
CHEMISTRY

1. Only dust – is there a sign of life?

- Determine whether the sample of dust originates from a living source (plant or animal) or a non-living source (rocks or sand).
- Design a test for the dust sample that exploits a physical property of organic molecules from living sources, using only the equipment provided.

2. Finding the 'rate expression' for the reaction between iodine and tin

- Find the 'rate expression' for the reaction between tin and a solution of iodine in methylbenzene (toluene), which produces tin (IV) iodide:

  $Sn(s) + 2I_2(s) \rightarrow SnI_4(s)$

Iodine and tin(IV) iodide are both soluble in methylbenzene.

- Design and carry out an experiment to measure the rate at which tin reacts with iodine.

THE ROYAL
SOCIETY OF
CHEMISTRY

3. Discover the properties of an element

- Use your scientific knowledge and skills to find out what you can about the properties of the element that you have been given. Can you suggest what it might be?

4. The hunt for vitamin C; the effect of cooking processes on the vitamin C content of cabbage

▼ Although vegetables in our diet are a source of vitamin C it is easily lost during cooking. How does the vitamin C content of cabbage vary with the method of cooking?

The amount of dissolved vitamin C can be determined by a titration using a solution of the dye 2,6-dichlorophenolindophenol. This is a dye which is blue when dissolved in water, red in acid conditions and is reduced by ascorbic acid (vitamin C) to a colourless compound.

An outline of a method for estimating vitamin C is available.

5. Nobili's rings

Nobili's rings are beautiful coloured concentric rings which are seen on the surface of metals under certain conditions during electrolysis.

This phenomenon was discovered in Italy by Leopold Nobili in 1827 who investigated the effect by using different metals and alloys as the electrodes and varying the salt solution in which they were immersed.

▼ Find out as much as you can about this phenomenon by setting up the electrochemical cell using the procedure described below.

Procedure

Prepare a saturated solution of lead ethanoate. Filter the solution and pour the filtrate into a flat shallow dish and place a flat piece of nickel foil or nickel-plated metal in the bottom of the dish. The plate is connected to the positive side of a DC supply via a switch and the negative electrode is a needle which is clamped above and very close to the plate. Do not allow a connection to be made between the electrodes.

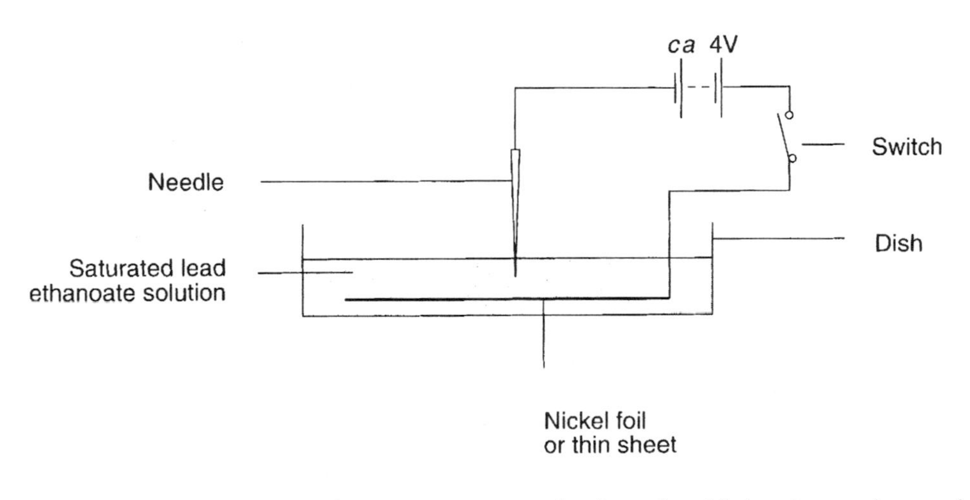

When the switch is closed concentric rings of colour should develop and spread outwards. For the best effects do not let the current flow for too long.

Immediately after the circuit has been disconnected remove the plate, rinse it in deionised water and allow it to dry. During the reaction lead will be deposited on the needle, and this should not be allowed to fall on to the plate.

6. From milk to curds and whey – which enzyme?

▼ Find out which of the three enzymes provided – rennet, genetically engineered chymosin and a fungal enzyme – is the best for coagulating milk to make curds and whey.

Traditionally rennet essence, obtained from the stomachs of calves or adult cows, is used for this process but biotechnology has made other enzymes available. The curds are used to make cheese, which is an enormous sector of the food industry. The whey can be treated to produce a sweet syrup that is used in preparing many different sorts of food.

7. Any glucose?

▼ Design a biosensor to detect glucose.

Glucose oxidase breaks down glucose in the presence of oxygen into hydrogen peroxide and gluconic acid. Horseradish peroxidase catalyses the oxidation of potassium iodide by hydrogen peroxide (formed by the action of glucose oxidase) to iodine, which is brown. The intensity of this brown colour can be taken as a measure of the amount of iodine present.

Safety note

8. Degrees of acidity

▼ Determine which of the two solutions labelled A and B is 0.1 mol dm^{-3} hydrochloric acid and which is 0.1 mol dm^{-3} sulphuric acid by using the simple equipment provided.

The only other solution available is 0.5 mol dm^{-3} aqueous sodium hydroxide. Indicator solutions should not be used.

9. Franklin's teaspoon of oil

Some remarkable outdoor experiments were carried out in the 1770s by Benjamin Franklin who was staying in London representing America in affairs of state. In 1776 he was one of the signatories of the Declaration of Independence. He was a man of genius who was both an international statesman and a scientist. To this day he is remembered for flying a kite in a thunderstorm to show that thunderclouds are electrified. He and some of his friends were interested in the effect of oil on the surface of water. The following extracts are taken from letters read to the Royal Society in 1774.

"...But recollecting what I had formerly read in Pliny, I resolved to make some experiment of the effect of oil on water, when I should have opportunity...

...At length being at Clapham where there is, on the common, a large pond, which I observed to be one day very rough with the wind, I fetched out a cruet of oil, and dropt a little of it on the water. I saw it spread itself with surprising swiftness upon the surface; but the effect of smoothing the waves was not produced; for I had applied it first on the leeward side of the pond, where the waves were largest, and the wind drove my oil back upon the shore. I then went to the windward side, where they began to form; and there the oil, though not more than a teaspoonful, produced an instant calm over a space several yards square, which spread amazingly, and extending itself gradually till it reached the leeside, making all that quarter of the pond, perhaps half an acre, as smooth as a looking-glass.

After this, I contrived to take with me, whenever I went into the country, a little oil in the upper hollow joint of my bamboo cane, with which I might repeat the experiment as opportunity should offer; and I found it constantly to succeed.

In these experiments, one circumstance struck me with particular surprise. This was the sudden, wide, and forcible spreading of a drop of oil on the face of the water, which I do not know that anybody has hitherto considered. If a drop of oil is put on a polished marble table, or on a looking-glass that lies horizontally; the drop remains in its place spreading very little. But when put on water it spreads instantly many feet round, becoming so thin as to produce the prismatic colours, for a considerable space and beyond them so much thinner as to be invisible, except in its effect of smoothing waves at a much greater distance..."

▼ Find out what you can about the dimensions of a molecule of olive oil by making calculations using the information contained in the historical account above. The capacity of 18th century teaspoons has been estimated to be *ca* 2 cm^3 and olive oil is mainly triolein (1,2,3-Tri-*cis*-9-octadecenoylglycerol).

Esso

Clapham pond before a drop of oil is added

Clapham pond after a drop of oil is added

10. Which solution is which?

The seven aqueous solutions, labelled A–G, are: copper(II) sulphate, dilute copper(II) chloride, concentrated copper(II) chloride, nickel(II) sulphate, barium chloride, potassium chloride and potassium iodide.

▼ Without using any other reagents identify which is which.

11. Are these jelly babies natural?

Introduction

Jelly babies are as attractive to look at as they are to eat because dyes in the translucent jelly give them a mouth-watering appearance. Traditional jelly babies contain synthetic dyes and nowadays the use of these dyes is closely controlled and monitored. These permitted colours have been thoroughly tested, but nevertheless many people prefer to buy 'natural' jelly babies that contain only natural dyes, because they believe dyes from nature must be less of a health hazard.

▼ A selection of jelly babies is provided. Some of them are coloured with synthetic food dyes while others are more fashionable with the manufacturers claiming that they contain no artificial colours. Which samples are 'natural' and which 'artificial'?

You should run paper chromatograms of all your extracts and compare them against reference samples of colours. All of the artificial colours should give clearly visible spots.

Method for extracting synthetic food dyes using wool

1. Dissolve three jelly babies of the same colour by heating in 100 cm^3 of deionised water.
2. Cut the wool into 20 cm lengths and pre-treat the wool by boiling it for 3 minutes in 0.05 mol dm^{-3} NaOH but do not exceed this time. Rinse the treated wool twice with deionised water, boil in deionised water for five minutes and then rinse twice more with deionised water.
   This process ensures that all the basic groups on the fibres have OH^-(aq) ions associated with them and that any excess of alkali is removed.
3. Boil 35 cm^3 of the jelly baby solution with a 20 cm length of wool for 5 minutes. During this time the wool should become coloured and the solution colourless as the anions of the dye(s) exchange for the OH^-(aq) ions. The wool should then be removed and rinsed with deionised water.
4. Boil the wool with 25 cm^3 of 2 mol dm^{-3} ammonia solution for 10 minutes. During this time the wool should lose its colour to the solution as the OH^-(aq) ions exchange for the anions of the dye.
5. Remove the wool and boil the extract to concentrate it to at least half its original volume. If the extract is still alkaline, acidify it with HCl before chromatography.

12. A transient red colour: the aqueous chemistry between iron(III) ions and sulphur oxoanions

If you bubble sulphur dioxide gas into an aqueous solution of iron(III) ions, a chemical reaction takes place: the iron(III) is reduced to iron(II). During this reaction a transient red colour is observed which is also seen when aqueous solutions of sulphur based salts are added to aqueous iron(III) ions.

▼ Which of the aqueous solutions provided give the transient red colour with aqueous iron(III) ions? By considering the structure of the sulphur compounds suggest an explanation for the red colour.

13. An analysis of coloured sands

Introduction

The pale yellow or brown colour of a sandy beach is due to the presence of iron compounds, mainly iron(III) oxide. Sand is made up of small fragments of rock; the most common mineral in sand is quartz (silica) with small amounts of mica and feldspar. Interestingly, pure silica sand is white. The iron(III) oxide usually forms a thin film or skin around each individual sand grain and it can cause the sand to be yellow-brown or red. There are many fine gradations and differences in hue and these are produced by varying concentrations of the iron(III) oxide colouring material In addition to this, variations in colour occur in sands of different grain sizes with the same content of colouring material. The iron(III) oxide is on the outside of the sand grains, hence it is relatively easy to remove.

▼ Find out which of the sand samples contains the most iron.

Possible procedures

Isolating the iron(III) oxide

The removal of the iron(III) oxide is the first practical stage in a chemical analysis of the sand. You have a choice of methods. The notes below are based on methods which are known to work; it is up to you to find the best method. An appropriate sample size for each method is 1g.

(i) Sublimation method

A boiling tube with a cold finger condenser is used. A little sand is mixed thoroughly with an approximately equal volume of solid ammonium chloride. The mixture is heated in the boiling tube when the sand will lose most of its colour and a yellow-stained sublimate will be obtained on the surface of the cold-finger and on the walls of the boiling tube. The iron is now in the sublimate as soluble iron(III) chloride.

(ii) Acid extraction method

An alternative approach is to dissolve the iron(III) oxide from the surface of the grains by using an acid: 2 mol dm^{-3} hydrochloric acid will give good results (8 mol dm^{-3} sulphuric acid has also been used).

Once the iron(III) ions are in solution the problem is to find a way to compare the amounts extracted from different sand samples.

Esso

14. What is honey made of? The optical rotation of natural sugars

Honey has the property of rotating the plane of polarization of polarized light. Sugars, which are the major components of honey, are chiral compounds and like almost all naturally occurring carbohydrates the sugars present in honey are the D molecules. Each sugar, at a given concentration, affects the rotation by an amount that is characteristic of the sugar. The optical rotation of a mixture will depend on the relative proportions of the sugars present. Honey analysts used optical rotation as a method of sugar analysis for many years.

Bees gather nectar from flowers and concentrate it to make a supersaturated solution of glucose, fructose and sucrose. The bees add an enzyme, invertase, and this converts most of the sucrose (a disaccharide) in the nectar to glucose and fructose. The resulting mixture has, at hive temperature, a very high solubility in water. This has two advantages for the bees. The low water content (18%) of honey means that it is resistant to spoilage through fermentation and honey also represents a very dense store of energy, taking up a minimum of space in the hive.

▼ Measure the optical rotation of honey. Honey is a supersaturated solution of mostly D-glucose and D-fructose, with a small amount of sucrose. Find out how each of these substances is likely to rotate the plane of polarisation of polarized light. Are they dextrorotatory or laevorotatory? Which of the sugars is present in the highest concentration?

It is suggested that a solution is made by transferring 26 g of honey to a 100 cm^3 volumetric flask which is then diluted to volume with water.

15. A yellow solid

▼ The following preparation should produce a good yield of a rather unusual compound. Find out as much as you can about the compound by using the tests suggested below.

Preparation

Dissolve 10 g of copper(II) sulphate crystals in 50 cm^3 of warm water. In a separate beaker dissolve 18 g of sodium thiosulphate crystals in 30 cm^3 of warm water. Adjust the temperature of each solution to about 40°C, and then pour the solution of thiosulphate into the copper solution with continuous stirring. After a short time, a bright yellow solid begins to separate, and is fully deposited in about an hour. Filter off the solid using a filter-pump and Buchner funnel and wash the solid on the filter paper with cold water followed by propanone. Finally, allow the solid to dry in air at room temperature.

▼ Investigate small quantities of the solid as follows.

(i) Heat some of the powder in a dry test-tube and test any gas evolved with a piece of filter paper soaked in acidified potassium dichromate(VI) solution.

(ii) Warm a little of the powder slowly with dilute nitric(V) acid until boiling.

(iii) Put a little of the solid in a test-tube followed by 10 drops of iodine solution and shake.

(iv) Put a little of the solid in a test-tube followed by dilute ammonia solution. Leave to stand for some time.

(v) Repeat **(iv)** but also add a spatula of ammonium persulphate crystals.

16. Detecting diluted evidence: screening body fluids for drugs of abuse

Introduction

A false positive report from a drug test can wreck an athlete's career and for any individual it can be a disaster. Traces of drugs in blood samples or urine can only be present if the subject has ingested the material, but the substance is often present at very low concentrations. In this problem you are asked to develop a method of concentrating a very dilute solution of a common drug in a sample of 'horse urine'. The aim is to produce a solution sufficiently concentrated to give a clear positive result in a chemical test. The setting is a horseracing forensic laboratory but the problem of extracting a substance from a very dilute solution is a general one in chemistry.

The drugs given to a horse might be found in urine at a concentration of 1μg cm^{-3} or less, whereas other naturally occurring components will be present at much higher levels. In our problem a 'model' horse consumes the equivalent of one tablet of aspirin (acetyl salicylic acid) in a glass of water. Inside the horse's body the aspirin is hydrolysed to form the metabolite salicylic acid which passes into the bloodstream. The kidneys help a little, as they gradually remove the salicylic acid from the blood and concentrate it in the urine, but the concentration is too low to give a decisive result in a chemical test.

An artificial sample of horse urine containing traces of salicylic acid, derived from aspirin, is prepared first.

The concentration equivalent to one aspirin tablet in a glass of water can easily be detected by a simple reaction with iron(III) chloride when a deep purple colour is formed. By the time the acid appears in the urine it is about 50 times more dilute and is only just detectable by the test. A further complication is that the urine contains many other coloured compounds which may mask the faint purple colour.

▼ Devise and test a method of extracting and concentrating the salicylic acid.

Suggested procedure

A 500 cm^3 beaker is provided, containing a sample of horse 'urine'. It is actually water coloured yellow by the addition of a very small amount of tartrazine.

1. Dissolve 1 ASPRO CLEAR in 50 cm^3 water ('urine') in a beaker.
2. Pour 1 cm^3 into a test-tube and add 1 cm^3 of Trinder's reagent. This should give a yellow colour.
3. Add 2 cm^3 of 4 mol dm^{-3} HCl and add an antibumping granule to the remaining ASPRO CLEAR in the beaker. Boil the solution for about 10 minutes in the fume-cupboard. The hydrolysis that occurs under these conditions represents the reactions which take place inside the horse. Allow to cool. Some of the water will have evaporated; replace this from a wash bottle.
4. Transfer 1 cm^3 of this hydrolysed aspirin solution to a test-tube and add 1 cm^3 of Trinder's reagent. A deep purple colour will be produced.

Esso

5. Pour 50 cm^3 of 'urine' into a beaker and add 1 cm^3 of hydrolysed aspirin using a pipette and stir.

6. Pour 1 cm^3 into a test-tube and add 1 cm^3 of Trinder's reagent. The colour change should be very slight.

You now have a 'urine' sample which is known to contain salicylic acid. From this sample obtain a much more concentrated solution which will give a clear positive result with Trinder's reagent.

17. As sweet as? Detecting aspartame in a table-top sweetener

A sample of a table top sweetener is provided. One teaspoonful of this would sweeten a cup of tea.

▼ Find out whether or not this substance contains the sweetening agent aspartame. Aspartame is a synthetic sweetening agent which tastes very similar to sucrose but is about 200 times sweeter.

Aspartame is the methyl ester of a dipeptide containing the amino acids aspartic acid and phenylalanine. The chemical name is L-aspartyl-L-phenylalanine methyl ester. The diagram below shows its structure with the dotted lines dividing the structure into its component groups.

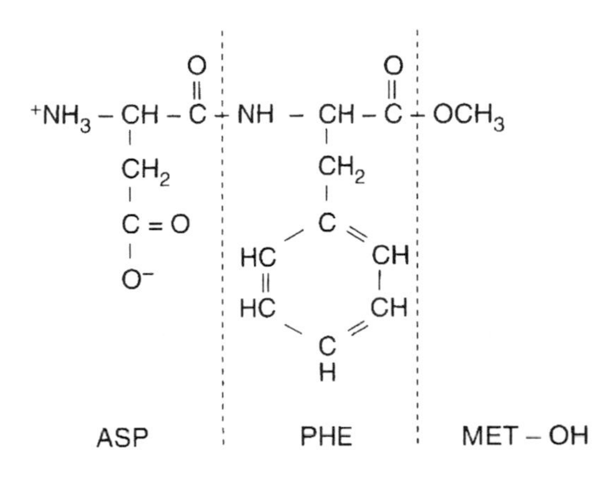

18. Liquid and solid water; the growth of ice crystals

On a cold clear night the Earth loses heat by radiation. The temperature of the air above a pond is several degrees below zero as a skin of ice begins to form, while the temperature of the liquid water below the ice is slightly above freezing. As the ice thickens the latent heat is conducted through the ice layer and radiated away into the night sky. Under these conditions the interface between the ice and the water is very nearly smooth.

(i) Imagine that the layer of ice is about 10 mm thick. Estimate from your experience how quickly the ice is thickening. Give your answer as a rate in mm s^{-1}.

(ii) Design an experiment to measure the rate at which ice freezes (or melts).

(iii) Carry out your experiment after having it checked to see if it is feasible.

(iv) It is quite easy to explain why ice floats on water. Can you explain why the pond does not freeze solid?

Data

Latent heat of water = 333 kJ kg^{-1}
Thermal conductivity of ice = 2.1 J s^{-1} m^{-1} K^{-1}
Density of ice = 0.92 kg m^{-3}

Esso

19. Vintage titrations: sulphur dioxide in wine

All wines contain sulphur dioxide. It is essential to add a certain amount to prevent the wine deteriorating and becoming unpalatable. The sulphur dioxide destroys bacteria that may cause unwanted secondary fermentation and it also acts as an antioxidant. However, if too much is added it will itself impart an unpleasant taste.

▼ Devise an analytical method to compare the amount of sulphur dioxide present in the samples of wine provided.

Free sulphur dioxide is present in wine as $SO_2(aq)$, $HSO_3^-(aq)$ and $SO_3^{2-}(aq)$. It is determined by titrating the acidified wine with iodine, using starch as an indicator.

$$SO_2(aq) + I_2(aq) + 2H_2O(l) \rightarrow 4H^+(aq) + SO_4^{2-}(aq) + 2I^-(aq)$$

Most of the sulphur dioxide in wine is combined with various acids and ketones, therefore to break down these compounds the wine is first treated with sodium hydroxide and acidified. The titration then gives the total sulphur dioxide.

20. Vintage titrations: tannin in wine

Tannin is a substance found naturally in red wines. Wines that contain a lot of tannin are described as 'full' and are said to have body; however, too much tannin makes the wine taste bitter. As wine matures the single tannin molecules slowly polymerise to give six or eight-unit tannins that are less bitter, mellowing the wine.

Wine-making is a more scientific business than the wine correspondents may lead you to believe. The tannin concentration can be measured by oxidation with potassium manganate(VII) using indigo carmine as the indicator. But the tannin concentrations determined by this method also include the pigments in wine. Like tannins, the pigments belong to a class of oxidisable compounds known as flavonoids.

Potassium manganate(VII) also oxidses the alcohol and other substances in the wine, including the indicator. To allow for the indicator and other oxidsable compounds a blank titration is carried out by gently boiling the wine to remove the alcohol. For the blank titration a sample of wine must be treated with activated charcoal to remove the tannins and the pigment.

▼ Determine the tannin concentration in the samples of wine. How do you account for the differences between red and white wine?

Esso

21. A taste for kilojoules: food calorimetry

The energy value of a food can be found by burning it and measuring the energy that is liberated as heat. In a living cell a complex series of enzymatic steps takes the place of the fire. The reaction pathways are different but, provided that the initial reactants and the final products are the same, an equivalent amount of energy is released. An organism uses this energy to support the processes of life.

People take food energy values into account when they make their everyday food choices. Recommended dietary allowances, once found only in textbooks on nutrition, are now used in daily life.

▼ Determine the food energy values of the foodstuffs provided. A purpose built food calorimeter attached to an oxygen cylinder will be needed. It is fairly easy to set fire to a peanut but many foods are less flammable and will only burn in an atmosphere enriched with oxygen. The calorimeter contains a known mass of water, a stirrer and a thermometer. The food to be burned is placed in a nickel crucible. Oxygen is led through the copper tube into the space around the crucible. The food is ignited, by an electrical device, and the rise in temperature of the water during combustion is measured. You will need to know the relationship between the increase in the temperature of the water and the amount of heat energy being produced.

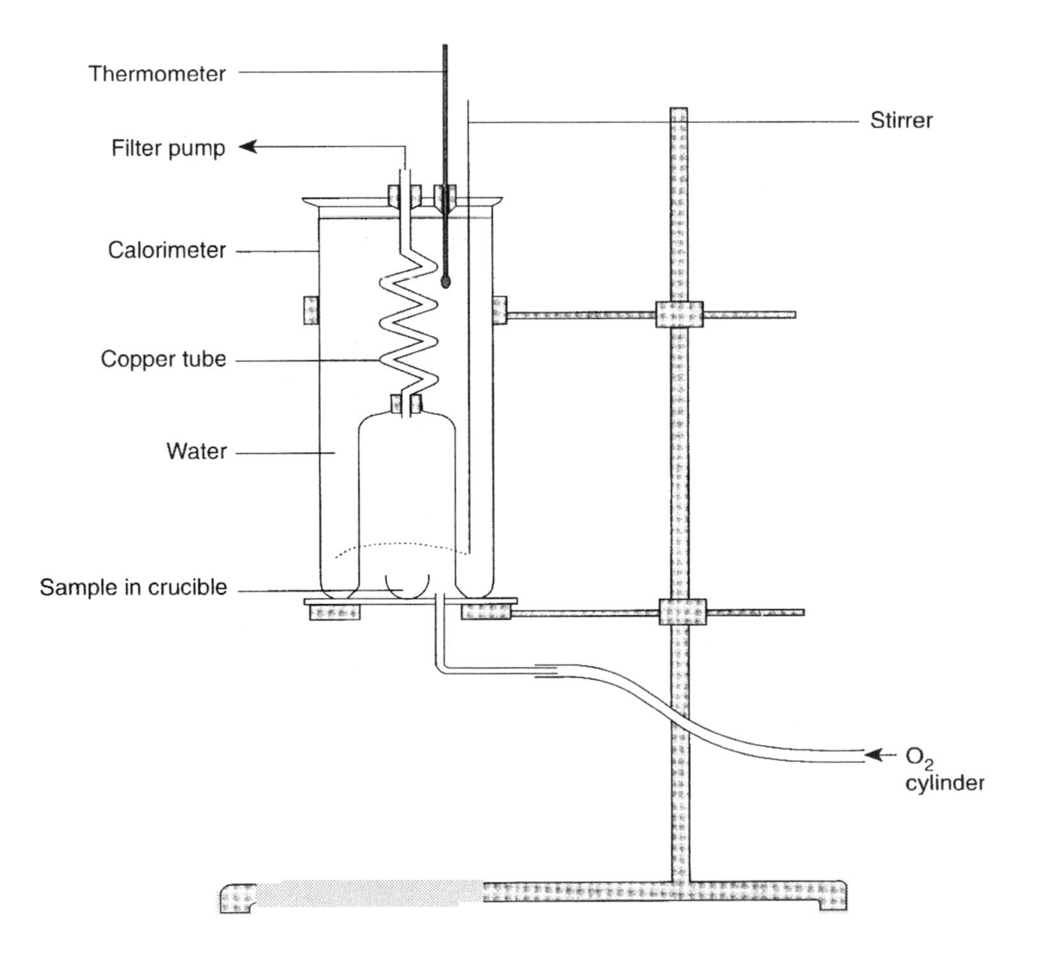

THE ROYAL SOCIETY OF CHEMISTRY

- Begin by testing familiar foods. Highly processed snack foods often contain unexpectedly large amounts of energy, usually as fats. You can check how well the apparatus works by comparing your results against known food values. You may decide to calibrate the calorimeter by using a food with a known energy value.

Esso

22. Theory v practice: do they compare?

The aim of this problem is to examine the reaction of calcium metal with water from both a theoretical and a practical viewpoint.

The reaction to be investigated is:

$$Ca(s) + 2H_2O(l) \rightarrow Ca(OH)_2(aq) + H_2(g)$$

Theoretical value

Draw up an energy cycle for the reaction. The cycle should include the electron affinity of the hydroxide radical (OH•) which has a value of -176.5 kJ mol^{-1}. The other enthalpy values should be those quoted in standard data books.

Experimental value

Devise an accurate method of measuring the temperature change during the reaction and use this value to determine the enthalpy values for the reaction. You must have your proposed method checked for safety before you start. The aim is to ensure that the most accurate experimental result is achieved.

23. Three isomeric alcohols

You are provided with three alcohols, labelled A, B and C, with isomeric structures that share the molecular formula $C_4H_{10}O$.

▼ Identify the three alcohols using standard laboratory tests.

24. Removing the vanadium

Alkenes (C=C) can be made to react to give oxiranes (epoxides) which have a three membered ring – *ie*

If a compound contains more than one double bond both can react in this way. In some cases, you can control which double bond will react if an alcohol (OH) group is present. A particularly useful reaction uses a vanadium complex as a catalyst and $(CH_3)_3COOH$ as the oxidising mixture – *eg*

$$(CH_3)_3C{-}OOH + CH_3CH_2 = CH - CH = CHCH_2OH \xrightarrow[\text{trichloromethane solvent}]{\text{VO (acac)}_2\text{ catalyst}} CH_3CH = CH - \overset{\overset{O}{\frown}}{CH - CH}CH_2OH + (CH_3)_3COH$$

The catalyst, vanadyl acetylacetonate [$VO(acac)_2$] is emerald green but turns red during the reaction, then returns to green afterwards. When the chemist carrying out the reaction tried to isolate the product a brown tar consisting of the crude product and the vanadium complex was obtained. The crude product was unstable to chromatography and the mixture could not be analysed by nuclear magnetic resonance (NMR) spectroscopy because the vanadium complex interfered with the signals and made the spectra useless.

The crude product was obtained by pouring the reaction mixture into a separating funnel, washing it with water (twice), [the $(CH_3)_3COH$ dissolves in water], separating off the organic layer and drying it over anhydrous magnesium sulphate. After filtering, the solvent was evaporated to leave the crude mixture.

▼ The chemist would like to take a NMR spectrum of the crude compound without the vanadium complex. Using your knowledge of vanadium chemistry devise a solution to the problem.

25. The Flatlandian Periodic Table

The periodic system of the elements in our three dimensional world is based on the four electron quantum numbers n = 1,2,3..., l = 0, 1..., n-1; m_l = 0, ±1, ±2 ..., ±l; and m_s = ±½. Let us move to Flatlandia. It is a two dimensional world where the periodic system of the elements is based on three electron quantum numbers: n = 1,2,3...; m = 0, ±1, ±2..., ±(n-1); and m_s = ±½. m plays the combined role of l and m_l of the three dimensional worlds (*ie* s,p,d,... levels are related to m). The following tasks and the basic principles relate to this two dimensional Flatlandia where the chemical and physical experience obtained from our common three dimensional world are applicable.

a) Draw the first four periods of the Flatlandian Periodic Table of the elements. Use the atomic number (Z) as the symbol of the element. Number the elements according to their nuclear charge. Give the electron configuration of each element.

b) Draw the hybrid orbitals of the elements with n=2. Which element is the basis for organic chemistry in Flatlandia? Give the Flatlandian analogues for ethane, ethene and cyclohexane. What kind of aromatic ring compounds are possible in Flatlandia?

c) Which rules in Flatlandia correspond to the octet and 18-electron rules in the three dimensional world?

d) Predict graphically the trends in the first ionisation energies of the Flatlandian elements with n=2. Show graphically how the electronegativities of the elements increase in the Flatlandian Period Table.

e) Draw the molecular orbital energy diagrams of the neutral homonuclear diatomic molecules of the elements with n=2. Which of these molecules are stable in Flatlandia?

f) Consider simple binary compounds of the elements (n=2) with the lightest element (Z=1). Draw their Lewis structures, predict geometries and propose analogues for them in the three dimensional world.

g) Consider elements with n<3. Propose an analogue and write the chemical symbol from our three dimensional world for each Flatlandian element. On the basis of this chemical and physical analogy predict which two dimensional elements are solid, liquid or gas at the normal pressure and temperature.

26. Which sodium salt is which?

The five sodium salts, labelled A-E, are: sodium sulphate; sodium sulphite; sodium thiosulphate; sodium metabisulphite; and sodium persulphate.

▼ Using laboratory tests identify which one is which and explain how you arrive at your answer.

27. What is the smallest amount that you can smell?

- ▼ Investigate the smallest amount of substance provided that can be smelt.

28. Which gas is which?

In the beaker are test-tubes containing different gases. The gases are carbon dioxide, dinitrogen oxide, oxygen, chlorine and hydrogen.

You may remove a test-tube only once and when you do so you must identify the gas immediately.

▼ Which tube contains which gas?

29. Identify the metal

▼ In the four blue solutions, labelled A–D, are transition metal complexes. All of the metals are in the first row of the Periodic Table. By using the reagents supplied identify the metals and, if possible, their complexes.

30. Which white salt?

▼ The four white solids are sodium carbonate, sodium hydrogencarbonate, sodium ethan-1,2-dioate (oxalate) and a mixture of all three. Carry out tests to identify each solid.

31. Design a pocket handwarmer

- Design a rechargeable handwarmer which, when it is needed, can be set off to provide a steady supply of heat at a comfortable temperature. This will be achieved by using an exothermic energy change.
- Produce a mock-up or experimental model of your device by using readily available materials.

32. A hot dinner from a can

▼ Produce a design for a self-heating can. The aim is to heat the food contained in a standard size tin to 65°C and to maintain it at this temperature for up to 60 minutes.

The can is intended for use on a camping expedition. It should be convenient to carry and easy to use; and the food will be eaten straight out of the can.

▼ Make a prototype for demonstration at the end of the session.

33. A chemical stop-clock: iodine clock reaction

If equal quantities of the two colourless solutions A and B are mixed together the mixture will remain colourless for a few seconds and then suddenly change to a deep blue colour. The time delay varies with concentration of the solutions.

▼ Find the dilutions that will give a time delay of exactly 70 s. Produce 20 cm^3 samples of diluted solutions of A and B which, when mixed, will enable the required time to be measured.

34. Lifting an egg by a thread

Cellulose fibres can be made by extrusion from a solution of dissolved cellulose.

Add about 10 g of copper carbonate to about 100 cm^3 of 880 ammonia solution in a beaker, until no more dissolves. After a while decant off the blue solution. Then stir in gently between 1 and 1.5 g of finely shredded cellulose, taking care not to fold in air, until the blue solution has the consistency of a gel. Cellulose fibres are reformed by extruding the solution into 1 mol dm^{-3} sulphuric acid solution. The fibres must be washed with water after their colour has faded.

▼ Make some cellulose fibres that will lift an egg.

35. Number of reactions

- How many chemical reactions can you identify by using the solution of copper(II) sulphate and solution A? Identify A.

36. Gas volume

- ▼ Construct apparatus to measure the molar volume of hydrogen produced from the reaction of magnesium with vinegar.

37. Blanching: what is the most effective method?

Fresh vegetables contain enzymes which, in time, help breakdown plant tissue, leading to 'off' flavouring. Freezing vegetables slows down the process but does not stop it.

To keep vegetables for a long time they are often blanched. This involves immersing the vegetables in boiling water for long enough to inactivate the enzymes without unduly softening or discolouring the vegetables.

The enzymes that are usually used as an indication of sufficient blanching are catalase and peroxidase (depending on the vegetable). The presence of peroxidase can be confirmed if a brown coloration is observed within 10 s of adding 3 drops of 0.5% hydrogen peroxide and 3 drops of 1% guaiacol solution to the same cool sample. (It is recommended that the samples are cooled in fresh water each time.)

▼ What is the most effective method of blanching sprouts?

38. Cool it

- By using the samples of citric acid and bicarbonate of soda (sodium hydrogencarbonate) provided devise a method of reducing the temperature of water to 6.5 °C when required. The temperature should be reached 1 minute after the reaction starts.

THE ROYAL SOCIETY OF CHEMISTRY

39. Liesegang rings: how the tiger got its stripes

▼ A set of instructions is provided which will enable you to produce the phenomena known as Liesegang rings. When you have seen for yourself some of the beautiful effects that can be produced, design an experiment to investigate the ring formation. There is no universally accepted theory to explain this phenomenon so it is up to you to frame hypotheses and to design experiments to test your ideas.

Two chemicals which, in solution, form a sparingly soluble precipitate are required. A gel containing one of the chemicals is prepared first. An aqueous solution of the other chemical is then placed in contact with the gel and after a while periodic precipitation patterns appear in the gel.

The patterns of rings or bands were named after Dr Raphael Liesegang who was the first scientist to make a systematic study of the phenomenon. Until 25 years ago they might have been classed as a scientific curiosity although they were studied in depth by physical chemists interested in certain specialised areas. Liesegang described the rings as rhythmic precipitation patterns occurring within a gel. It is now recognised that they are an example of a chemical reaction that oscillates in time and space. The spontaneous appearance of patterns is remarkable because the system is self-organising – the motions of billions of molecules have synchronised to create patterns in time and space. Liesegang thought that the banded effects seen in rocks, on the wings of butterflies and the skins of animals might have a similar origin to that of the effects he observed.

In 1957 some workers made and photographed in colour about a 100 experiments in silica gels. Thirty years later they compared the appearance of the gels with that in the original photographs. Recently a school in Kent planned to investigate the effect of microgravity on Liesegang rings by designing an experiment to be sent into space on board a US space shuttle. It probably won't be possible for you to make such ambitious plans but there are many other interesting lines of investigation.

Sets of instructions are provided which should enable you to produce very clear Liesegang ring systems in a preliminary experiment.

Preliminary instructions

Preparation of a gel containing potassium dichromate(VI) covered by silver nitrate solution.

(a) Weigh out 2.5 g of gelatin and 0.025 g of potassium dichromate(VI). Add 50 cm^3 of deionised water. Heat gently, with stirring, until the solution is clear. Pour into test-tubes of various diameters so that the tubes are about two-thirds full. Cover the test-tube with parafilm and leave the solution to set.

(b) Weigh out 0.85 g of silver nitrate and dissolve in 10 cm^3 deionised water. About 1 cm^3 of this solution is poured on top of the gel. The test-tube is then covered with parafilm and left undisturbed.

Over the next few days the formation of bands of colour will be observed. The experiment may be modified by pouring a small amount of the gelatin-potassium

dichromate(VI) solution onto a glass slide or into a crystallising dish and, after it has set, dropping on a very small amount of silver nitrate solution. Concentric rings will be seen to develop.

Preparation of a gel containing cobalt(II) chloride covered by 880 ammonia solution.

Weigh out 1.5 g gelatin and 2.5 g cobalt(II) chloride and add 50 cm^3 deionised water. Heat gently with stirring until the solution is clear. Pour the solution into a test-tube (25 x 150 mm) until it is about two-thirds full. Cover the tube with parafilm and leave undisturbed until a gel is formed. Carefully pour 880 ammonia solution on top of the gel until the tube is almost full. Cover the tube with parafilm. The tube should be left to stand for a few days when bands will be observed to form.

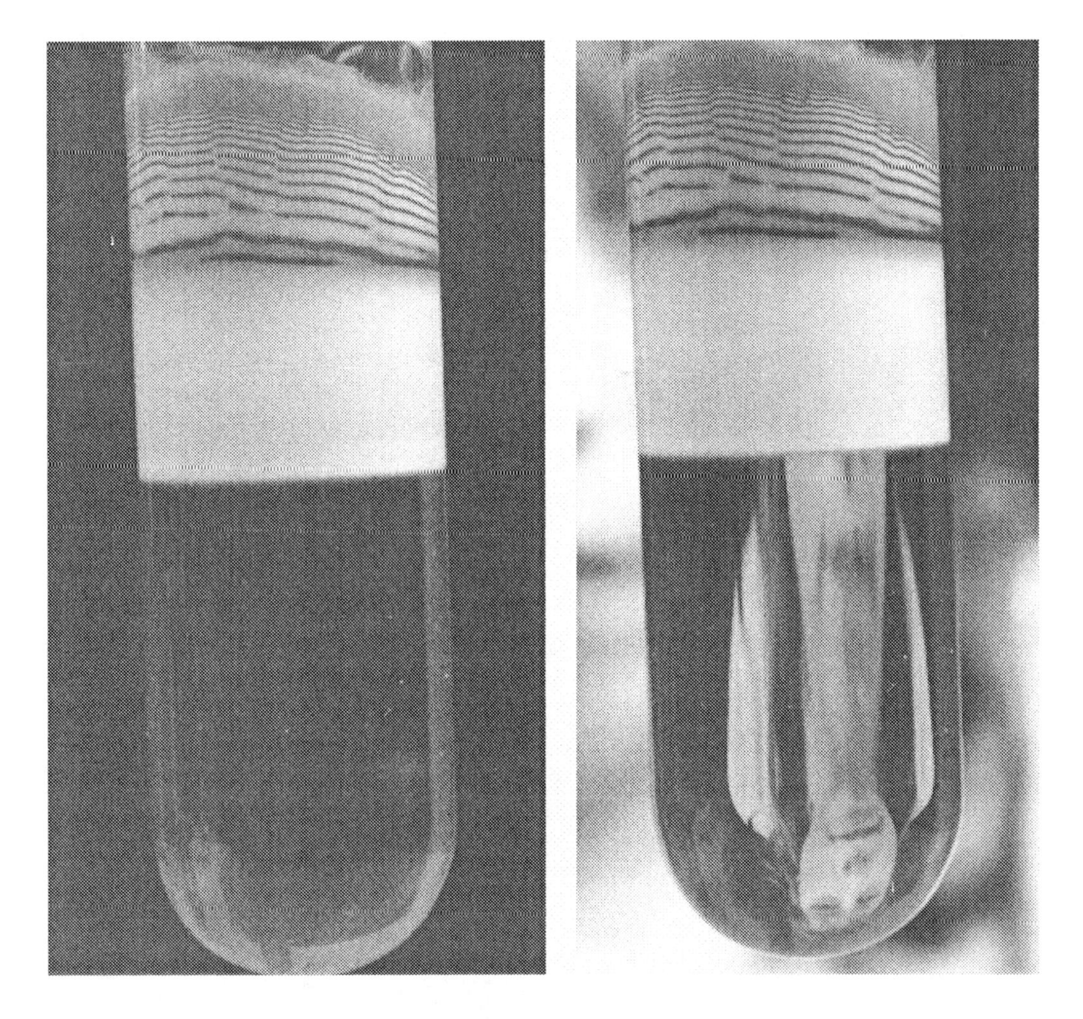

Photographs of Liesegang rings

THE ROYAL SOCIETY OF CHEMISTRY

40. The candle in the bell-jar

Introduction

In 1990 Stephen Pople wrote an article in the *Education Guardian* describing the hidden dangers of insulation in modern homes. Although energy can be saved by cutting out draughts it is dangerous to seal a room completely. A building needs to have a steady supply of fresh air because the oxygen used up by fires and people has to be replaced. A version of an experiment, known in old textbooks as 'The candle in the bell-jar' was given in the article as a way of showing children how oxygen can be used up in a room that is sealed too tightly.

In the traditional experiment a candle is burnt in a bell-jar over water. The candle burns for a short time and is then extinguished. As the candle burns the water level rises, apparently showing that a fraction of the air is being used up. A reduction in volume of the gas by about one-fifth may be observed and this is sometimes claimed to be a neat 'proof' that air is 20% oxygen.

Stephen Pople claimed that:

the rise in water level shows how much oxygen is used up during combustion

His claim was challenged in a letter:

In the demonstration with the "suffocating candle" the explanation given is that the oxygen is used up and the water rises to take its place. A word equation for this combustion would be: Hydrocarbon in candle wax (solid) and oxygen (gas) becomes carbon dioxide (gas) and hydrogen oxide or water (liquid).

As oxygen is used up then carbon dioxide is produced. The demonstration is puzzling because it does not account for where the carbon dioxide produced has gone. The given explanation encourages children to think of combustion as the using up of oxygen with no thought about the products of combustion. If Stephen Pople tries the demonstration again and carefully observes the water level, he will notice that the water does not start to rise as the beaker seals the floating candle. This suggests that the change in water level may be more to do with the physics of cooling gases than the chemistry of combustion.

This letter was followed by a letter from a second person, a research chemist. He did not accept the explanation in terms of cooling gases and wrote:

There may well be an effect such as he [the author] describes which may contribute to the observed volume change. There is however one fatal flaw in his reasoning for rejecting the chemical explanation, which he dismisses on the grounds that an equal quantity of gaseous carbon dioxide is formed and hence there is no volume change. If we think more carefully on the products of combustion we will see that his word equation is incomplete. Paraffin wax is a solid hydrocarbon with a chain length of 20 – 30 and the molecular structure approximates to CH_2. The chemical equation for the combustion reaction is therefore:

$2CH_2$(solid) + $3O_2$(gas) $\rightarrow$ $2CO_2$(gas) + $2H_2O$(liquid)

In other words, three volumes of oxygen are replaced by two volumes of carbon dioxide, representing an overall reduction in gas volume. I believe the experiment is still a valuable educational aid.

- Find out as much as you can about what is going on when a candle burns in an enclosed space to investigate the claims made above. Try repeating the original experiment and then develop and modify it.

41. Move an Oxo cube at great speed

- Design and make a device to move an Oxo cube as fast as possible over a 1 m flat surface. The energy source used must be the reaction between 1 level teaspoon of bicarbonate of soda (sodium hydrogencarbonate) and 3 level teaspoons of citric acid.

As far as possible, the device is to be constructed from 'junk' materials. Your final device must be loaded with chemicals, and it must be ready to start when you are told to do so.

Esso

42. Move a heavy object

- Design and make a device to move the heaviest possible object at least 10 cm on a flat surface. The energy source used must be the reaction between 1 level teaspoon of bicarbonate of soda (sodium hydrogencarbonate) and 3 level teaspoons of citric acid.

As far as possible, the device is to be constructed from 'junk' materials. Your final device must be loaded with chemicals, and you must be ready to start it when told to do so.

43. Test the gas

▼ Design a piece of apparatus to allow you to obtain a sample of the gas without removing the gas jar from the water and demonstrate that the gas is carbon dioxide.

44. What makes the candle go out?

- If a lighted candle is placed inside a gas jar everyone knows that it will go out. What happens if two or more candles are placed inside?

45. Warning device

- Design a warning device that will give out a signal when activated. The device must be powered by the chemical reaction between 3 level teaspoons of citric acid and 1 level teaspoon of bicarbonate of soda (sodium hydrogencarbonate). You are to use only the amounts of chemicals supplied. A fresh supply will be given to you when judging takes place.

The device must give a clear warning, and be activated only when you are told to do so.

Esso

46. An elementary problem

▼ Five chemical companies are awaiting chemicals from suppliers and they will be delivered on different days. From the information given, which company has ordered which chemical, which process is involved, and on what week day are they to be delivered?

Company E has ordered a chemical for the Solvay process while the sodium hydroxide is for company B. Both of these orders must be delivered before those for company D and the company that operates the Kroll process.

The vanadium(V) oxide is needed for the Contact process while the chemical for the Bayer process is needed before Friday. The magnesium is not required by company A.

The chemical for the Haber process, which is not sodium hydroxide is to be delivered on Tuesday while the calcium carbonate that was not ordered by company D is to be delivered on Wednesday.

Potassium oxide is one chemical on order and the chemical for company C will be delivered on Thursday.

47. Making ice

▼ Which makes ice faster, hot or cold water?

48. Fizzy drinks

▼ If you shake a can of fizzy drink before opening it the drink froths over the side of the can. If you leave the can to stand after shaking the drink froths less, if at all, when opened. Why?

49. After meal puzzle

An ice cube floats on water in a glass tumbler or a cup. Apart from a piece of string the only items available are those that you would expect to find set out on a meal table.

▼ The puzzle is to remove the ice with the help of the string. Ingenuity is required; using cutlery or tying the string around the cube are forbidden.

50. A chemically powered boat: a bubble boat race

- Design and make a boat propelled by the reaction between 1 teaspoon of bicarbonate of soda (sodium hydrogencarbonate) and 3 teaspoons of citric acid. The fuel can be carried on the boat or the gas can be generated separately and stored in a balloon. The boat is to be constructed from cheap, readily available materials and must be powered by bubbles. The aim is to design the boat that travels the furthest distance.

THE ROYAL
SOCIETY OF
CHEMISTRY

Esso

Esso

1. Only dust – is there a sign of life?

Time

2 h on the day. A hand-out, including a list of the contents of the kits, can be given to groups the day before .

Level

A-level, Higher Grade or equivalent.

Curriculum links

Optical activity, stereospecificity of terrestrial life.

Group size

3 – 4.

Materials and equipment

Materials per group

Dust:

- sand
- vermiculite
- D-fructose

Each sample should contain *ca* 5 g D-fructose.
(Alternatively, amino acids can be used.)

- For calibrating polarimeter: 10 g D-fructose, deionised water.

Equipment per group

Items from the junk list (pXX). Miscellaneous items including:

- sealing wax
- rubber bands
- rubber tubing
- glass tubing and rods (various)
- protractor
- paper
- test-tubes
- test-tube holder and rack
- beakers (various)
- weighing bottles (various, flat bottomed)
- filter funnel
- filter paper
- 25 cm^3 and 100 cm^3 measuring cylinders

- petri dishes
- pasteur pipettes
- optical filters (coloured, polarizing, clear, diffuse)
- clamps and stands
- corks
- copper wire
- light source
- wash bottle
- mirrors (small)
- Bunsen burner
- plastic gloves
- safety glasses.

Safety

Eye protection must be worn.

Risk assessment

A risk assessment must be carried out for this activity.

Commentary

Some students may need help in remembering that the building blocks of proteins and carbohydrates – amino acids and sugars – are chiral and therefore stereospecific. They may also need help in recalling how a polarimeter works. However, the instruction to test a physical property of organic compounds from living sources, plus the presence of optical filters in the equipment, prompted most students to test for optical activity by constructing a polarimeter. This problem has been used successfully in competitions.

Possible approach

The hand-outs, including a list of the contents of the kits, can be given to groups the previous day. The main challenge must be seen as the construction of an arrangement capable of giving a reliable estimate of the optical activity in the sample.

A possible design for a polarimeter is sketched below; a way of measuring the angle of rotation must be devised.

Protractor
Polaroid sheet
Cardboard tube enclosing solution
Polaroid sheet
Light

Extension work

If this method was used to analyse dust from a meteoric crater for signs of extra terrestrial life, what assumptions would the students have to make to analyse their results?

Evaluation

This problem was set as a competition and marks were awarded as follows.

1. Marks were awarded for the construction of the polarimeter.
2. Marks were awarded for use and demonstration of optical activity in the dust.
3. Marks were deducted for hints that were given.

Acknowledgement

This problem is based on a suggestion by John Liggat and originated from a competition set in the Chemistry Department of the University of Glasgow in 1986.

2. Finding the 'rate expression' for the reaction between iodine and tin

Time

2–3 h, but it could be used as an extended project. Each 'run', using a given concentration of iodine takes 30 minutes.

Level

A-level, Higher Grade or equivalent (or beyond).

Curriculum links

This experiment could be used as an introduction to reaction kinetics.

Group size

2 – 4. It is also suitable as a demonstration experiment, when the data could be shared.

Materials and equipment

Materials per group

- tin foil (0.5–1.0 mm thick) as a rectangle 10 mm x 25 mm or 3 g tin cast in the form of discs or cylinders
- propanol-1-ol for cleaning the surface of the tin
- 20 cm^3 of 10% w/v solution of iodine in methylbenzene
- methylbenzene for diluting the solution above to give 10 cm^3 samples of lower concentrations. (The suggested range is 10% to *ca* 2% iodine in methylbenzene.)

Equipment per group

- ignition tube or boiling tube for casting tin
- tall narrow 25 cm^3 beaker or weighing pot without lid
- balance with underweighing facility reading to ±0.001 g
- fine emery paper
- safety glasses
- gloves
- access to a fume cupboard.

Safety

Eye protection must be worn.

Methylbenzene must be used in a fume cupboard.

Risk assessment

A risk assessment must be carried out for this activity.

Commentary

Tin reacts relatively easily at room temperature with a solution of iodine in methylbenzene. The experiment should prove instructive because the students can observe the progress of the reaction directly. A balance is arranged so that it can weigh a disc or cylinder of tin hanging suspended in an iodine solution. The balance reading falls steadily with time. The procedure is repeated with different concentrations of iodine; the results should show first order behaviour.

Rate of loss of tin = $k[I_2]^1$

It is most important that the surface of the tin be cleaned immediately before each run. It should be rubbed carefully with fine emery paper lubricated with water, rinsed with deionised water and finally rinsed with propan-1-ol. After cleaning the surface should not be touched .

This experiment is described in an earlier Royal Society of Chemistry publication.[1] Since then the design of balances available for use in schools and colleges has changed and it has become possible to connect a computer to the balance and use it to process the data.

Procedure

The simplest option is to use tin foil. This needs to be thick enough to allow the surface to be thoroughly cleaned with emery paper before each experiment. Although the heavier gauge foil is more expensive, it is possible to clean and to reuse a sample many times. If the reaction does not work, the most probable reason is that the tin is not clean enough. The reaction with insufficiently cleaned tin takes 12 h, but with clean tin readings can be taken every minute.

The alternative method is to cast tin into a suitable shape. There are two ways of doing this:

(i) Some tin is melted in an ignition tube,[2] and the drops are pushed together with a glass rod. When the tin has solidified, the tube is cracked by immersing it in cold water. A small hole is drilled through the end of the tin bar so that it can be suspended from the balance arm by a monofilament nylon thread. The surface of the tin should be smoothed with fine emery paper and rinsed with alcohol.

(ii) A bar of tin, about 15 mm in diameter, is cast in a glass boiling tube.[3] The tin bar is cut into discs about 1 mm thick using a small hacksaw or on a lathe. Small holes are drilled through the discs, which are then smoothed with fine emery paper and rinsed in alcohol. Monofilament nylon thread is then used to suspend the discs.

It is necessary to raise the balance to use the underweighing facility as shown in the diagram below.

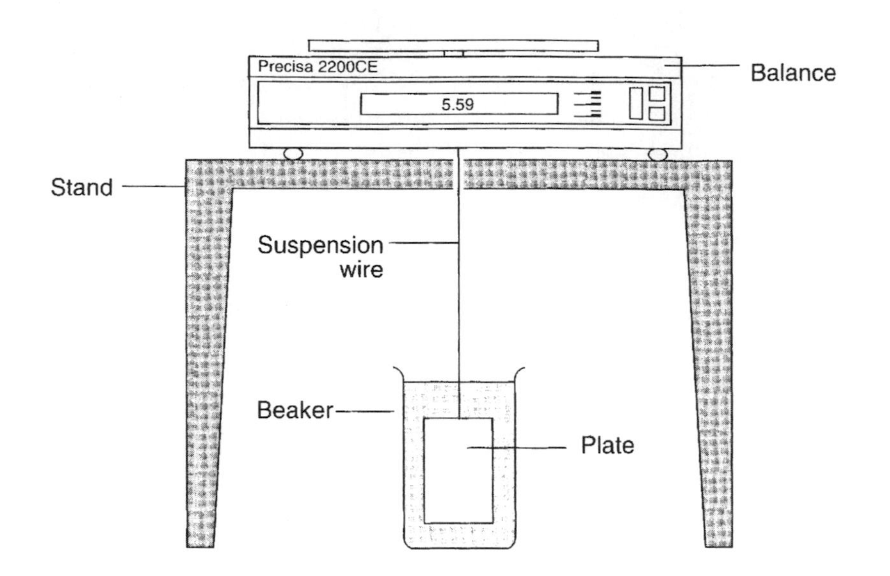

A 10% w/v solution of iodine in methylbenzene is recommended for the first reaction. This can be diluted to give varying concentrations down to 2%. It should be possible to obtain a set of readings within 20 minutes.

Extension

Interfacing a digital balance to a PC or to a BBC computer has become a standard procedure.[4] It is particularly simple in the case of BBC machines and this experiment lends itself to this approach; the experimental data can then be printed out in the form of graphs.

The experiment could be the basis of more detailed studies on heterogeneous systems.[5] If tin foil is available the effect of varying the surface area could be studied.

References

1. B. E. Dawson, C. L. Mason and P. Mason, *Reaction kinetics, a resource for teachers*. London: RSC, 1981.

2. *Nuffield advanced science sourcebook, physical science*. B. E. Dawson (ed). London: Penguin, 1974.

3. E. J. F. Davies and A. F. Trotman-Dickenson, *J. Chem. Educ.*, 1966, **43**, 483.

4. R. Edwards, *Interfacing chemistry experiments*, London: RSC, 1993.

5. E. J. F. Davies and A. F. Trotman-Dickenson, *J. Chem. Soc.* 1947, 736.

THE ROYAL SOCIETY OF CHEMISTRY

3. Discover the properties of an element

Time

1–2 h or longer for full investigation of electrical properties.

Level

A-level, Higher Grade or equivalent.

Curriculum links

Ability to interpret the chemical and physical properties of an element in terms of its position in the Periodic Table. A knowledge of the mechanism of conduction in a semiconducting material may be an advantage.

Group size

2.

Materials and equipment

Materials per group

- Silicon, available in lumps

  If the students want to make quantitative measurements of the effect of temperature on the resistivity, pieces of the purest grade available should be used. The silicon should be kept as dry as possible.

- Bench top reagents: *eg* nitric acid, sulphuric acid, hydrochloric acid, aqueous potassium hydroxide, aqueous sodium hydroxide.

Equipment per group

- test-tubes
- Bunsen burner
- equipment for measuring electrical resistance
- means of attaching copper leads to the silicon lumps
- safety glasses
- heating device – *eg* hot air blower or hair dryer. Some form of water bath may be needed.

For quantitative measurements on the effect of temperature on resistivity students may need electrically conducting epoxy resin which is available from suppliers of electronic components.

Safety

Eye protection must be worn.

Risk assessment

A risk assessment must be carried out for this activity.

Commentary

John Elmsley's book[1] would be an excellent resource for students attempting to solve this problem. The resistivity quoted in this book is low and corresponds to a 'metallurgical' grade.

This problem has been used successfully with A-level and International Baccalaureate students. It has also been used with a group of HND students who used the semiconducting properties of the element as the basis of a long term project.

The silicon should prove to be unreactive towards acids but will dissolve in hot alkalis liberating hydrogen. At red heat silicon reacts with water vapour[2] to form SiO_2. The melting point is too high for normal laboratory equipment. The electrical resistance should be found to be high compared with that of a metal but low compared with that of a non-metal, hence the term semiconductor.

When the element is heated the resistance is found to decrease and this behaviour is characteristic of a semiconductor. Heat energy is transferred to electrons in the valence band, and the increase in energy enables them to jump up into the conducting band. The number of conducting electrons and the number of positive holes increase so that more charge carriers are available and the electrical current increases for a given potential difference.

The resistivity reflects the purity level and the effect will only be found in purer grades of semiconductors. Hence a piece of pure silicon will have a higher resistance than the lumps. Pearson and Bardeen reached 100 ohm cm material in 1949 and were the first to describe silicon's semiconducting properties – *eg* an activation energy of 1.115 eV at room temperature.[3] Although Berzelius named and described the chemical element silicon in 1822, its semiconducting properties were not discovered until much later.

Extension

It should be possible to estimate the energy gap between the valence band and the conducting band by examining the dependence of the electrical resistance on temperature.

References

1. J. Elmsley, *The elements*, 2nd edn. Oxford: OUP, 1991.
2. *Handbook of the physicochemical properties of the elements* (trans), G.V. Samsonov (ed). London: Osbourne, 1968.
3. G. L. Pearson and J. Bardeen, *J. Phys. Rev.*, 1949, **75**, 865.

4. The hunt for vitamin C; the effect of cooking processes on the vitamin C content of cabbage

Time

2 h.

Level

A-Level, Higher Grade or equivalent.

Curriculum links

Mole calculations. Vitamins, enzymes.

Group size

2 – 4.

Materials and equipment

Materials per group

- 100 g green cabbage
- 1 dm^3 of a solution containing 5% orthophosphoric acid (H_3PO_4)
- 100 cm^3 of aqueous 2, 6-dichlorophenolindophenol (dcpip) (0.4 g dm^{-3})
- 75 cm^3 of ascorbic acid (0.20 g dm^{-3}) in 5% orthophosphoric acid solution
- Deionised water, boiled to remove dissolved oxygen which could otherwise interfere with the results of the vitamin C determination.

Equipment per group

- filter funnel
- muslin or glass wool for filtration
- 25 cm^3 pipette with safety filler
- 50 cm^3 burette
- 250 cm^3 conical flask
- 500 cm^3 measuring cylinder
- 250 cm^3 beaker
- Bunsen burner, tripod and gauze
- liquidiser, blender (or large pestle and mortar)
- safety glasses.

Although not essential, during trialling, some institutions used a fume cupboard to reduce the smell of over-cooked cabbage!

Safety

Eye protection must be worn.

Risk assessment

A risk assessment must be carried out for this activity.

Commentary

If cabbage is not cooked carefully the ascorbic acid (vitamin C) is broken down by the enzyme ascorbic acid oxidase. The secret of preparing nutritious cabbage is to plunge it rapidly into boiling water which inactivates this enzyme. Nevertheless, more than 50% of the vitamin C will be leached out into the water and therefore lost unless the liquid is used as an ingredient in another item on the menu.

This problem is based on an experiment described in *Nuffield advanced science chemistry.*[1] To determine the vitamin C content in uncooked cabbage, it is essential to have an efficient blender which grinds the raw material into a fine slurry. Otherwise it will be impossible to extract all of the ascorbic acid. Once the cabbage has been softened by cooking the blending process is quite easy. Unless a very efficient blender is available it is suggested that the first sample of cabbage is cooked by the 'nutritionally sound' method described below. The vitamin C content of the solid and the liquid can then be determined separately. It should be found that the sum of the quantities determined agrees with the value for raw cabbage quoted in the standard reference books on food and nutrition *ie* green cabbage[2] contains about 50 mg per 100 g.

A second sample can be cooked 'badly' by putting it into cold water and slowly bringing it up to the boil. It could be cooked for too long and left to stand. The vitamin C content in this sample will be lower.

Details of a recommended procedure for carrying out the analysis are given below, but during trialling more experienced students found a shorter version of these instructions sufficient. The challenge in this problem lies in devising a satisfactory sampling technique and controlling the cooking time, temperature and method. It is therefore recommended that more able groups of students are left to work out their own method of sampling while others are given the full instructions.

A method of calculating the results is suggested below.

Procedure

Preparation of 2,6-dichlorophenolindophenol solution

2,6-dichlorophenolindophenol (dcpip) is a dye which is blue when dissolved in water, is red in acid conditions, and is reduced to a colourless compound by ascorbic acid. Dissolve 0.4 g of dcpip in 200 cm^3 of hot deionised water, filter the solution, and make the volume up to 1 dm^3. The dye does not keep well and should be stored in a cool dark place.

Standardisation of the 2,6-dichlorophenolindophenol solution

The solution should be standardised because it is not possible to make it up accurately.

By using a pipette, with safety filler, transfer 25.0 cm^3 of standard ascorbic acid solution (0.20 g dm^{-3} vitamin C) to a conical flask and titrate rapidly with the dye solution from a burette. As the dye is run in the deep blue colour of the dye is discharged to give a colourless solution. The end point is taken to be when the pink colouration, due to the dye, persists for 10 s. A blank titration using 25.0 cm^3 of 5% orthophosphoric acid solution must be carried out to the same end point.

$$F = \frac{\text{volume of standard vitamin C solution x concentration of vitamin C/mg dm}^{-3}}{\text{(standardisation titre – blank titre) x 1000}}$$

These quantities are then used to calculate the dye factor (F)

F = mg of vitamin C equivalent to 1 cm^3 of dye solution

Estimation of vitamin C in a sample of cabbage

Cut up the cabbage as if preparing it for a meal. Weigh out 50 g to an accuracy of ± 0.5 g. Put the cabbage into 100 cm^3 of briskly boiling deionised water and simmer the cabbage for 10 minutes. Some of the liquid will have evaporated. Pour off the hot water and measure its volume (V_c).

At this point one person should measure out 250 cm^3 of 5% orthophosphoric acid and add this to the cooking water then transfer 25 cm^3 of this solution to a conical flask, using a pipette, and titrate it with the dye.

Meanwhile someone else should quickly liquidise the cooked cabbage. They should then remove it from the liquidizer and add 250 cm^3 of 5 % orthophosphoric acid solution and the mixture stirred. It should then be weighed (mass M_c).
About $^1/_{12}$ of the mixture should be removed and weighed (mass m_c). This fraction should be filtered through the muslin or glass wool and the filtrate and washings should be made up to about 25 cm^3 (the exact volume at this stage is not necessary for the calculation).

Calculation

(i) Liquidised cabbage

Mass treated = 50 g

Mass sampled for titration $= \frac{m_c}{M_c} \times 50$ g

Volume of dye titre = V cm^3

Dye factor = F mg cm^{-3}

50 g of sample contains $V \times F \times \frac{M_c}{m_c}$ mg vitamin C

100 g of sample contains $V \times F \times \frac{M_c}{m_c} \times 2$ mg vitamin C

(ii) Cabbage water

Original quantity of cabbage = 50 g

Proportion sampled for titration $= \frac{25}{V_c + 250}$

Volume of dye titre = V cm^3

Dye factor = F mg cm^{-3}

50 g of sample contains $V \times F \times \frac{V_c + 250}{25}$ mg vitamin C

100 g of sample contains $\frac{V \times F \times V_c + 250 \times 2}{25}$ mg vitamin C

Extension

Experiments could be carried out to find the effect of other cooking methods on the vitamin C content of cabbage – *eg* samples could be stir-fried, microwaved, or steamed.

The techniques described above could also be used to find out how freezing affects the nutritional value of cabbage. Freezing slows down deterioration because of the inactivation of the enzymes. However, if the blanching process is carried out correctly, before freezing, the enzymes should be destroyed.

References

1. *Nuffield advanced science chemistry: food science, A special study*. London: Longmans, 1970.
2. B. Holland *et al*, McCance and Widdowson's *The composition of foods*, 5th edn. London: RSC, 1992.

5. Nobili's rings

Time

1.5 h or longer.

Level

A-level, Higher Grade or equivalent.

Curriculum links

Simple electrochemistry. One line of investigation depends on understanding interference phenomena in light and therefore requires a background in physics.

Group size

2 – 4.

Materials and equipment

Materials per group

- nickel foil
- aluminium foil
- other metals, plated metals
- standard needles, very fine needles such as hyperdermic needles, wire – these can be bent to form the negative electrode
- deionised water
- lead ethanoate
- other salts as indicated in the *Commentary* below.

Students are invited to use their ingenuity in this investigation and they should be encouraged to suggest possible materials to be used as electrodes.

Equipment per group

- DC power supply or battery
- switch
- connecting wires
- wash bottles
- glass dish, such as a petri dish
- safety glasses
- gloves.

Safety

Eye protection must be worn.
Skin contact with the materials should be avoided.

Risk assessment

A risk assessment must be carried out for this activity.

Commentary

Professor Iolo Williams took Nobili's Rings as the basis for a challenge for young chemists set to pupils in South Wales schools in 1977 to celebrate the centenary of the Royal Institute of Chemistry. This problem is a version of the original challenge.

Nobili[1] discovered the phenomenon and studied it using electrodes made from silver, gold, platinum, tin, bismuth and brass. He also varied the composition of the electrolyte. The effects seem to have been particularly striking when he used a solution of lead ethanoate. The brilliance of the colours is improved if the metallic surface is highly polished. When the switch is closed concentric rings of colour should develop and spread outwards. For the best effects, do not let the current flow for too long.

Another reference,[2] mentions the beautiful effects that can be achieved by using negative electrodes of copper wire bent into different shapes, such as a cross or a star.

The colours are due to interference effects in the light reflected from the metal, but the students should be left to discover this for themselves. The tints vary with the thickness of the thin films of oxide that build up on the electrode surface. The colours that appear on the surface of a specimen of heat-treated steel have a similar origin.

Some results using different anodes and electrolytes are as follows:

Anode	Electrolyte	Notes
nickel	lead ethanoate	excellent rings
nickel	lead nitrate	excellent rings
nickel	manganese sulphate	excellent rings
polished copper	lead ethanoate	good rings
polished copper	copper sulphate	fair rings
copper	lead ethanoate	poor rings
tin	lead ethanoate	very poor rings
nickel	copper sulphate	no rings
nickel	sodium hydrogencarbonate	no rings
lead	lead ethanoate	no rings
zinc	sodium chloride	no rings

The variables to experiment with are: concentration of the electrolyte, nature of the electrolyte, temperature, voltage, anode-cathode distance, shape of cathode, degree of polish of the metal.

References

1. L. Nobili, *Ann. Chim. Phys.*, 1827, **84**, 280.

2. *School Sci. Rev.*, 1935, **64**, 495.

Acknowledgement

This actvity is based on the idea suggested by Professor Iolo Williams.

THE ROYAL SOCIETY OF CHEMISTRY

6. From milk to curds and whey – which enzyme?

Time

1 h.
Extension 1 h.

Level

A-level, Higher Grade or equivalent.

Curriculum links

Use of enzymes in cheese making. Catalytic action of enzyme.

Group size

2–3.

Materials and equipment

Materials per group

- 40 cm^3 of whole pasteurised milk ('silver top')
- 1 cm^3 of rennet essence (of bovine origin, containing both chymosin and pepsin)
- 1 cm^3 of 'Maxiren' (pure chymosin) from genetically engineered yeast
- 1 cm^3 of 'Rennilase' (microbial protease) from a fungus
- 1 cm^3 of deionised water.

For the Extension

Making a sweet syrup from the whey

- 2 cm^3 of lactase enzyme (Novo 'Lactozym®')
- 8 cm^3 of a 2% (w/v) sodium alginate solution (Note: sodium alginate is not readily soluble and requires both warm water and stirring to dissolve)
- 100 cm^3 of a 1.5% (w/v) calcium chloride solution, Diastix™ (available from pharmacies) or other glucose detector.

Maxiren, Rennilase and Novo Lactozym® are available from the National Centre for Biotechnology Education, Department of Microbiology, University of Reading, Whiteknights, PO Box 228, Reading RG6 2AJ. (Telephone 01734 873743.)

Equipment per group

- 10 cm^3 plastic syringes without needles
- 4 boiling tubes or sample bottles
- 4 glass rods
- thermometer

THE ROYAL
SOCIETY OF
CHEMISTRY

- stop-watch
- access to water bath maintained at 37°C
- Safety glasses

For the extension

- Small piece (about 1 cm^2) of nylon gauze *eg* net curtain
- 10 cm^3 plastic syringe (without needle)
- 10 cm length of 4 mm diameter aquarium airline tubing to fit syringe
- aquarium airline tap or adjustable laboratory tubing clip (Hoffman clip)
- retort stand, boss and clamp
- 2 small beakers (*ca* 100 cm^3) or disposable plastic cups
- tea strainer.

Safety

Safety guidelines are given in the Teacher's notes of the National Dairy Council Publication.[1] (General guidance will be found in *Microbiology – An HMI guide for schools and non-advanced further education.* London: HMSO, 1985.)

Eye protection must be worn. Unnecessary contact with the enzyme or inhalation of dust from dried-up enzyme spills should be avoided. In case of spillage or contact with the eyes, rinse by flushing with water.

Risk assessment

A risk assessment must be carried out for this activity.

Commentary

This investigation is based on an problem designed by Dean Madden and John Scholar of the National Centre for Biotechnology Education for a practical workshop at the Natural History Museum, London, in May 1992.[1] On this occasion the students worked in groups of three and each group was given a box containing all the necessary apparatus. The practical details are adapted from a National Dairy Council Publication.[2] If the extension is completed this problem will fit well with 'Any glucose?'.

Procedure

The aim of this experiment is to compare the milk-clotting abilities of the three enzymes. The students should be able to design the experiment and appreciate that they may need to warm the milk to 37°C.

The formation of a precipitate may then be observed within 5 – 15 minutes. It may be seen adhering to the sides of the tubes when they are gently rocked from side to side, or clinging to a glass rod dipped into the liquid.

The activities of the three enzymes should be found to vary markedly.

Extension

You can extend the problem by investigating the activity of the enzymes at different temperatures and at different pH values.

Making a sweet syrup from the whey

Whey contains lactose which can be split up by the enzyme lactase to the simpler

sugars glucose and galactose. These sugars are sweet and are the basis of a syrup that is used widely in the food industry. An interesting extension is therefore to ask the students to make this sweet syrup, containing glucose, from the whey by using an immobilised enzyme column (see below).

Practical details

The enzyme lactase converts the whey to a sweet syrup and this enzyme can be immobilised by trapping it in calcium alginate beads. The beads are packed into a small column, over which the whey is passed. Instructions for setting up the column and treating the whey are given below.

Immobilising the lactase enzyme and packing the column[3]

1. Draw up 2 cm^3 of enzyme in a syringe and transfer to a small beaker.
2. Using the same syringe, transfer 8 cm^3 of sodium alginate solution to the beaker and mix.
3. Draw the mixed solution into the syringe.
4. Pressing the plunger gently, add the mixture to the calcium chloride solution drop by drop.
5. Strain the beads.
6. Rinse with distilled water.
7. Pack the beads into the column.

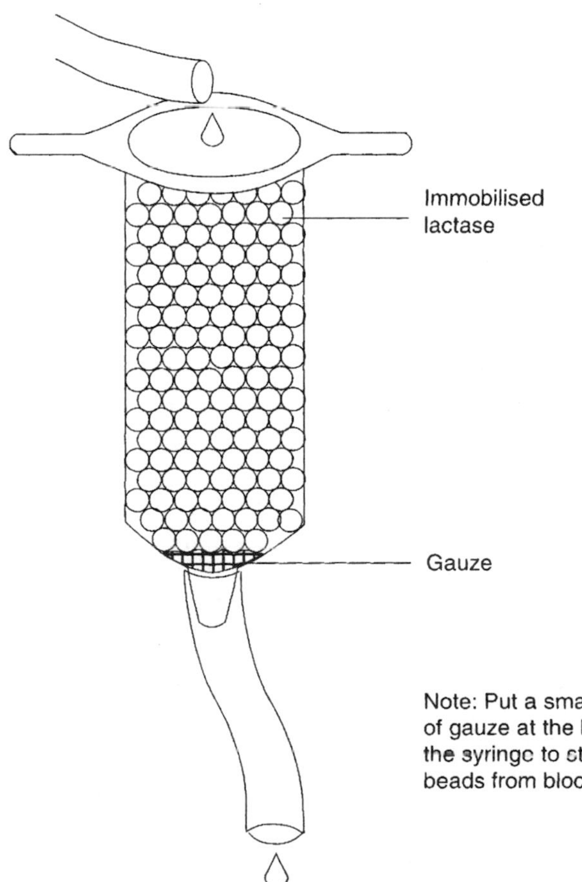

8. Treat the whey by pouring it through the column. The syrup coming out of the syringe should be found to contain glucose.

Tips

The tip of the syringe must not be allowed to come into contact with the calcium chloride solution. The immobilised enzyme beads should be allowed to harden for a few minutes before being separated from the liquid with a tea-strainer.

References

1. D. Madden, *National centre for biotechnology education Newsletter*, p12, Spring 1992.
2. *Dairy Biotechnology*, National Dairy Council 5/7 John Princes Street, London W1M 0AP.
3. D. Madden, *National centre for biotechnology education Newsletter*, p22, Spring 1992.

Acknowledgements

Dean Madden and John Schollar of the National Centre for Biotechnology gave advice on the development of this activity and the experimental procedures described were developed in their laboratories.

7. Any glucose?

Time

1 h.

Level

A-level, Higher Grade or equivalent.

Curriculum links

Catalytic action of enzymes. Simple carbohydrate chemistry.

Group size

2–3.

Materials and equipment

Materials per group

- a few drops of glucose oxidase/horseradish peroxidase mixture
- Fermcozyme 952 DM (available in 50 cm^3 quantities from Hughes and Hughes, Unit 1F, Lowmoor Industrial Estate, Tonedale, Wellington, Somerset, TA21 0AZ. Tel. 01823 660222)
- a few drops of 2% w/v potassium iodide solution
- solutions of glucose at various concentrations to test the sensor for the comparison with a commercial product Diastix™ (available from pharmacies).

Equipment per group

- dropping pipettes or 1 cm^3 plastic syringes (without needles) for dispensing liquids
- filter paper
- scissors
- forceps for handling enzyme-soaked paper
- safety glasses.

Safety

Safety guidelines are given in the Teacher's notes of the National Dairy Council Publication.[1] (General guidance will be found in *Microbiology – An HMI guide for schools and non-advanced further education. HMSO. London: 1985.)*

Eye protection must be worn. Unnecessary contact with the enzyme or inhalation of dust from dried-up enzyme spills should be avoided. In case of spillage or contact with the eyes, rinse by flushing with water .

Risk assessment

A risk assessment must be carried out for this activity.

Commentary

The students can be given the worksheet, the reagents and no other information. Test strips for detecting the presence of glucose in urine can be purchased at pharmacies (*eg* Diastix™, which works on a principle similar to that described on the student sheet). The students may be interested to compare their glucose detectors with commercially available products such as Diastix™.

This investigation is based on an activity designed for a workshop at the Natural History Museum, London, in May 1992.[1] On this occasion the students worked in groups of three and each group was given a box containing all the necessary apparatus. The practical details are adapted from a National Dairy Council Publication.[2] This is a useful adjunct to 'From milk to curds to whey – which enzymes?'

Practical details[3]

Special care must be taken to ensure that cross-contamination of the enzyme mixture, potassium iodide and sugar solutions does not occur. Separate, clearly marked syringes or pipettes should be used for dispensing each liquid.

A possible approach is as follows.

1. Cut out a small square of filter paper, roughly 10 mm x 10 mm.
2. Place a drop of potassium iodide solution on the paper, and allow to dry slightly, then add a drop of the enzyme mixture to the paper.
3. Add a drop of the glucose solutions to the paper and note the colour change; this may take a few minutes, as oxygen has to diffuse into the solutions.

Extension

During trialling several institutions extended the problem by getting students to investigate how sensitive the sensor was to the concentration of glucose. The activity can be extended to test the specificity of the sensor by using sugars other than glucose.

References

1. D. Madden, *National centre for biotechnology education Newsletter*, p12, Spring 1992.
2. *Dairy Biotechnology*, National Dairy Council 5/7 John Princes Street, London W1M 0AP.
3. D. Madden, *National centre for biotechnology education Newsletter*, p22, Spring 1992.

Acknowledgements

Dean Madden and John Schollar of the National Centre for Biotechnology gave advice on the development of this activity and the experimental procedures described were developed in their laboratories.

8. Degrees of acidity

Time

1–1.5 h.

Level

A-level, Higher Grade or equivalent.

Curriculum links

Exothermic/endothermic reactions and neutralisation reactions. Ionic equations. Strong and weak acids.

Group size

2.

Materials and equipment

Materials per group

- 50 cm^3 of 0.1 mol dm^{-3} hydrochloric acid (solution A)
- 50 cm^3 of 0.1 mol dm^{-3} sulphuric acid (solution B)
- 100 cm^3 of 0.5 mol dm^{-3} sodium hydroxide.

For the Extension

- 50 cm^3 of 0.1 mol dm^{-3} phosphoric acid.

Equipment per group

- –5 to +50°C thermometer
- two 25 cm^3 measuring cylinders
- two 50 cm^3 measuring cylinders
- polystyrene cup
- stirring rod
- safety glasses.

Safety

Eye protection must be worn.

Risk assessment

A risk assessment must be carried out for this activity.

Commentary

The solution to this problem relies on students carrying out a neutralisation reaction which results in a temperature rise, and is an example of calorimetric analysis. By using a data book[1] it is possible to predict the temperature rise when suitable volumes of strong acid and alkali are mixed. For strong acids and alkalis the molar enthalpy of neutralisation is effectively constant at about –57 kJ mol^{-1}. During trialling

results close to this value were obtained especially if a cooling curve correction was used.

Extension

The students could investigate whether this method can distinguish between a solution of phosphoric acid and sulphuric acid.

The enthalpy of neutralisation[2] of weak acids or alkalis may be greater or smaller than –57 kJ mol^{-1}. For example

Acid	Alkali	ΔH/kJ mol^{-1}
HF	NaOH	–68.6
HCN	KOH	–11.7

Reference

1. *Revised Nuffield advanced science book of data.* London: Longman, 1988.
2. A. M. James and M. P. Lord, *Macmillan's chemical and physical data.* London: Macmillan, 1992.

9. Franklin's teaspoon of oil

Time

1 h.

Level

A-level, Higher Grade or equivalent.

Curriculum links

Bond lengths.

Group size

1–2.

Commentary

The students are invited to interpret a historical experiment based on a full and interesting account of Franklin's contribution to surface chemistry written by Professor Charles Giles[1] whose researches show that the site of the experiment was the Mount Pond on Clapham Common. Franklin's communications were published in the *Philosophical Transactions* of the Royal Society.[2]

Calculation

If it is assumed that the oil forms a monomolecular layer then the thickness of the film should correspond to the height of a molecule of triolein lying on the surface of the water. Franklin estimated that one teaspoonful of olive oil spread to cover half an acre of the pond's surface.

Half an acre = $2420\ \text{yd}^2 = 2420 \times 0.9144^2\ \text{m}^2$

One teaspoonful, say $2\ \text{cm}^3 = 2 \times 10^{-6}\ \text{m}^2$

Hence the thickness of the film when one teaspoonful covers one half acre is:

$(2 \times 10^{-6})/(2420 \times 0.9144^2) = 9.9 \times 10^{-10}\ \text{m}$

Extension

This experiment is the forerunner of the classical 'oil drop experiment'.[3] Although teachers will probably have met this, it may be unfamiliar to students who could be asked to repeat Franklin's experiment on a small scale. Designing an experiment in which the pond is replaced by water contained by the sides of a flat tin tray would provide several opportunities for problem solving.

Historical background

The earliest recorded observations of adding oil to water seem to date back to the 18th century BC in Babylon. Today this region is known as Iraq. The phenomena observed when oil was poured into a bowl of water were seen as omens for the future and descriptions of the phenomena and the events they foretold were inscribed in cuneiform on clay tablets.

In a more recent publication on the history of surface chemistry[4] Giles and his co-authors refer to the researches of Tabor[5] who describes the Babylonian texts on the spreading of oil on water.

Franklin's account stimulated further work on the European continent. This has been researched by Scott[6] who describes the incentive offered by Frans van Lelyveld.

There are many references in classical literature to the way in which oil forms a thin film on water. Pliny is mentioned at the beginning of the passage quoted. Franklin did not claim to have made a discovery, he designed what is probably the first scientific experiment on a thin oil film and made observations that he communicated to the scientific world of 18th century Europe.

Franklin's interpretation stopped short of a simple calculation that might have led him to speculate on the size of particles of matter. At that time scientists were more excited by the idea that it might be possible to use oil to calm waves in stormy seas. In 1775 a Dutchman, Frans van Lelyveld, offered a prize of 30 ducats for the best suggestion as to how this could be done.

A few years later Franklin fell out of favour with the British public because he enlisted France's help for the American cause in the War of Independence. The British press attacked him and it seems that his scientific achievements were also belittled. Franklin's researches in surface chemistry became disregarded by British scientists although several papers were written on the continent on the subject of wave-damping. Interest in surface chemistry did not revive in Great Britain until the end of the 19th century.

Two hundred years after Franklin's experiment the historical researches of another surface chemist, Professor Charles Giles, enabled him to identify the pond, which still existed on Clapham Common. He repeated Franklin's experiment taking photographs of the effect. The area which is 'as smooth as a looking glass' shows the extent of the oil .

References

1. C. H. Giles, *Chem. Ind.*, 1969, 1616.

2. B. Franklin, *Phil. Trans. R. Soc. London*, 1774, **64,** 445.

3. *Revised Nuffield physics teacher's guide years 1 and 2.* London: Longman, 1978.

4. C. H. Giles, S. D. Forrester, G. G. Roberts, *Langmuir Blodgett Films*, G. Roberts (ed). London: Plenum, 1990.

5. D. Tabor, *J. Colloid and Interface Sci.*, 1980, **75**, 240.

6. J. C. Scott, *History of technology, Vol. 3.*, A. R. Hall, and N. Smith (eds). London: Mansell, 1978.

Acknowledgements

This activity is based on a suggestion from Dr R. Aveyard, who kindly lent the Society the transparencies from Professor Giles' original photographs. Thanks are also due to Dr S. D. Forrester for his advice and encouragement.

10. Which solution is which?

Time

1 h.

Level

A-level, Higher Grade or equivalent and able younger students.

Curriculum links

Tests for chloride, sulphate and copper ions. Complexes formed by copper and nickel ions.

Group size

2–3.

Materials and equipment

Materials per group

- labelled bottles (A–F) of 0.1–0.2 mol dm^{-3} aqueous copper(II) sulphate, nickel(II) sulphate, barium chloride, potassium chloride and potassium iodide. The dilute copper(II) chloride should be made up so that it is blue and the concentrated copper(II) chloride so it appears green.

Equipment per group

- test-tubes and test-tube rack
- safety glasses.

Safety

Eye protection must be worn.

Risk assessment

A risk assessment must be carried out for this activity.

Commentary

At first sight some students feel that this problem is impossible, but with a little thought it can be solved. Essentially the solutions are used to 'identify' each other by a process of elimination.

Procedure

This is one suggestion.

(i) Mix the colourless solutions with the blue ones. The two blue solutions are either copper(II) sulphate or dilute copper(II) chloride. The observations should be:

	$CuSO_4$	$CuCl_2$
$BaCl_2$	white ppt	goes green
KCl	goes green	goes green
KI	white ppt/ violet ppt	white ppt/ violet ppt

This enables you to identify the five solutions.

(ii) Add the barium chloride to the green solutions. The observations should be:

	$NiSO_4$	$CuCl_2$
$BaCl_2$	white ppt	stays green

The two green solutions are identified.

Acknowledgement

The activity is based on an idea from Terry Allsop.

11. Are these jelly babies natural?

Time

1 h is needed to extract the dye from a sample of jelly babies of the same colour. It is suggested that each group prepares extracts of one or two different colours (depending on the time available). The extracts can then be shared round so that each group has a complete set of colours for chromatography.

1 h for paper chromatography.

Level

A-level, Higher Grade or equivalent.
Suitable for younger pupils as a practical exercise.

Curriculum links

Students with a background knowledge of colour chemistry could use this problem to develop their understanding.

Students should be aware of colour additives, E-numbers and paper chromatography.

Group size

2.

Materials and equipment

Materials per group

- packets of jelly babies may be shared between groups, each group making extracts of one or two colours. Some jelly babies should contain artificial dye colours (*eg* quinoline yellow E104, sunset yellow E110, ponceau (red) E124, green G E142, brilliant black BN (bluish violet) E151); and some jelly babies should contain 'natural' food dyes, available from Boots the Chemist, (*eg* curcumin E100, copper chlorophyll (green) E141, ß-carotene E160a, annatto extract E160b, beetroot red or betanin E162).
- white double knitting wool (must be 100% wool – not a mixture)
- 50 cm^3 of 0.05 mol dm^{-3} sodium hydroxide
- 50 cm^3 of 2 mol dm^{-3} ammonium hydroxide
- deionised water
- chromatography solvent (2% w/v NaCl in 50% ethanol/water)
- standard synthetic food colours (available from shops specialising in materials for icing cakes). They may be selected to correspond with the synthetic colours described on the packets of jelly babies
- safety glasses.

Equipment per group

- scissors
- 50 cm^3 and 100 cm^3 beakers
- Bunsen burner, tripod and gauze

- test-tubes and test-tube racks for the coloured extracts
- plastic forceps are useful for handling and rinsing the wool. Standard paper chromatography equipment.

Safety

Eye protection. If pure food dyes are used, it is essential that protective clothing is worn because they will also dye fabrics.

Risk assessment

A risk assessment must be carried out for this activity.

Commentary

There have been claims that azo dyes in foods cause hyperactivity in young children but the evidence is inconclusive. The controversy surrounding the use of colour additives is reviewed by Ben Selinger in his book *Chemistry in the Marketplace*[1] in which he explains some of the chemistry behind the physiological effects. Synthetic food dyes contain aromatic rings, mainly benzene and naphthalene. They are sometimes known as coal-tar dyes, named after the starting materials once used for their manufacture. To make the synthetic dyes soluble in water, one or more sulphonic acid (SO_3H) groups are attached.

In this problem a wool dyeing process is used to detect the presence of synthetic dyes. The synthetic dyes are adsorbed onto the wool because the SO_3H groups are attracted to positively charged groups on the wool fibre. They can then be stripped from the wool using a solution of ammonia. When the extracts are analysed by paper chromatography the dyes appear as coloured spots and can be identified if standard samples are available.

Natural dyes differ in their structural features (see appendix). They do not have SO_3H groups. Although the natural dyes may be partially removed from solution onto the wool during the dyeing process, they tend to remain adsorbed on the wool when the colour is stripped.

The wool dyeing method used to be a standard procedure in determining the colorants present in foods.[2] Today other active substrates, for example polyamide powder, can be used in place of wool because the extraction process isolates the colours in a pure form. This is desirable since the presence of impurities may well affect their chromatographic behaviour. The extraction process concentrates the colour and the chief concern of the analyst is usually to determine whether the colour of a foodstuff is a permitted colour. This means that the additive has to be positively identified by instrumental techniques, especially spectrophotometry.[3]

Paper chromatography is still preferred for certain types of analysis in the laboratories of companies that manufacture food dyes.

Procedure

Stage 1. Use of wool to extract the dyes

The method is given in the student's section of this activity.

Stage 2. Paper chromatography

This is carried out using standard techniques. The dyes present may be identified if pure samples are available.

Extension

A similar but more advanced version of this experiment has been described.[4] Wool is used to extract synthetic food dyes from a variety of foodstuffs and the dyes are then separated and identified using thin layer chromatography and ultraviolet/visible spectroscopy.

References

1. B. Sellinger, *Chemistry in the marketplace*, 4th edn. London : Harcourt Brace Jovanich, 1989.
2. *Separation and identification of food colours permitted by the colouring matters in food regulations 1957.* London: The Association of Public Analysts, 1960.
3. D. M. Marmion, *Handbook of US colorants*, 3rd edn. Chichester: J. Wiley, 1991.
4. E. A. Dixon and G. Renyk, *J. Chem. Educ.*, 1982, **59**, 67.

Acknowledgements

This activity is based on a suggestion from Susan Jackson. Danny White of Warner Jenkinson Europe provided help and advice during the preparation of this activity.

Appendix

Examples of synthetic food dyes

Quinoline Yellow E104 Dye Classification: quinophthalone

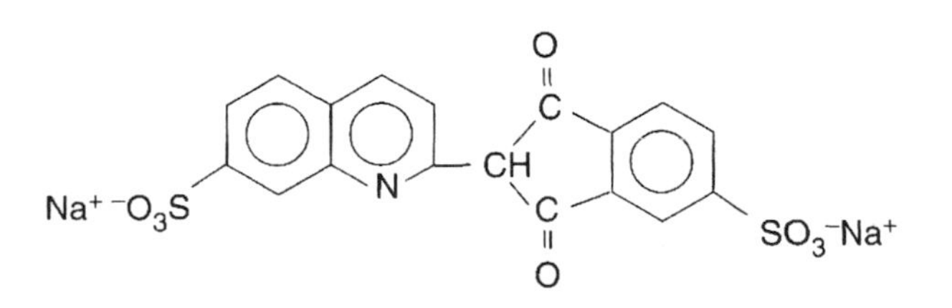

Sunset yellow E110 Dye Classification: azo

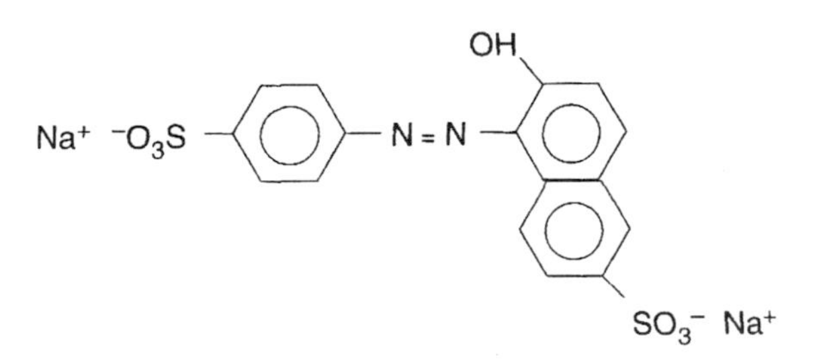

Esso

Ponceau (red) E124 Dye Classification: azo

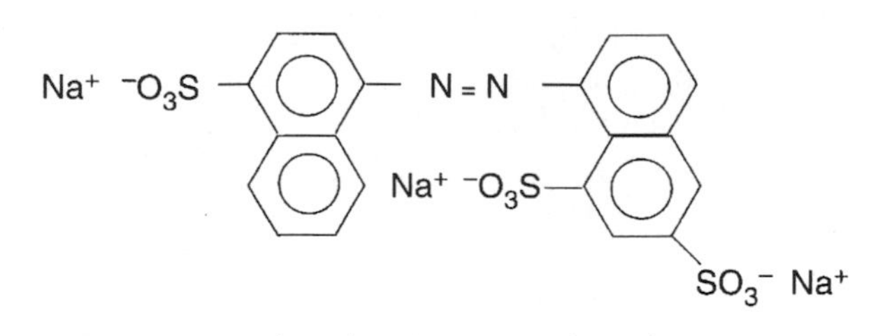

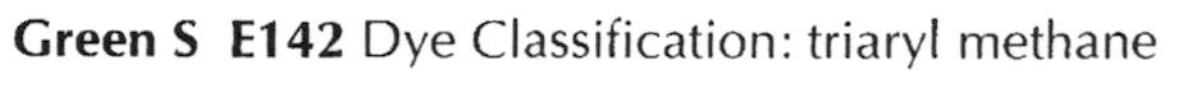

Green S E142 Dye Classification: triaryl methane

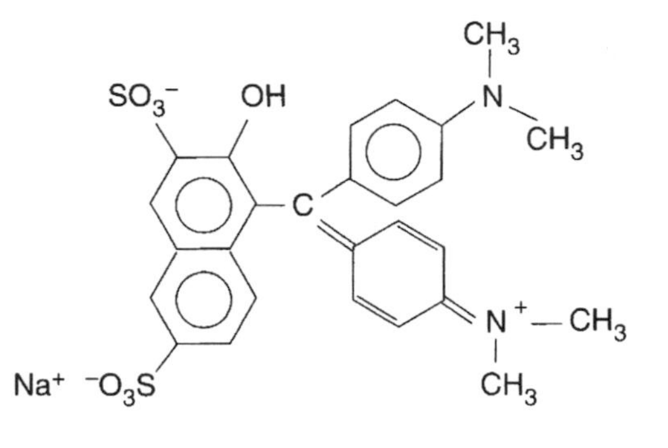

Brilliant black BN E151 Dye Classification: disazo

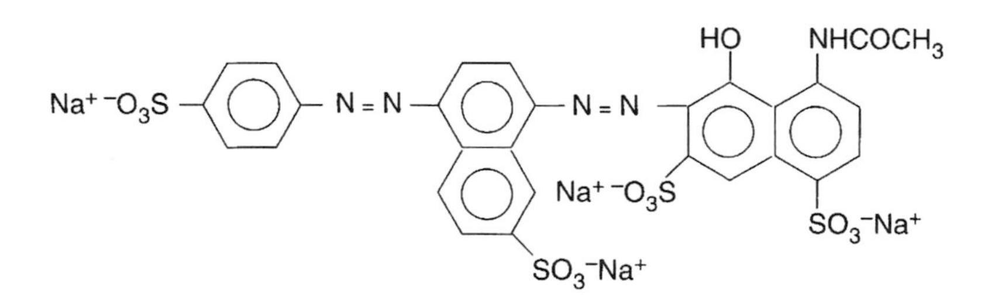

Examples of food dyes of natural origin

(These may be produced synthetically.)

β-carotene E160a

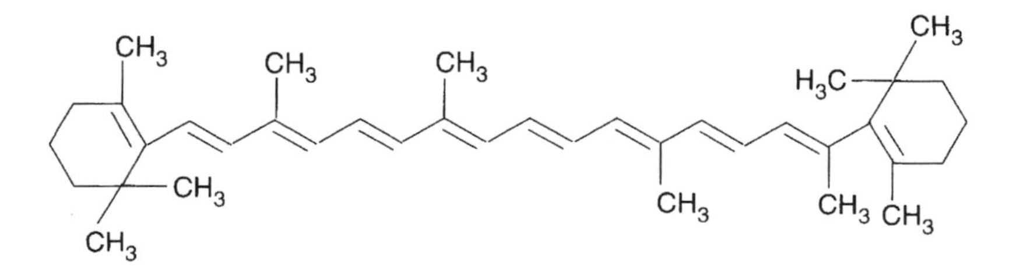

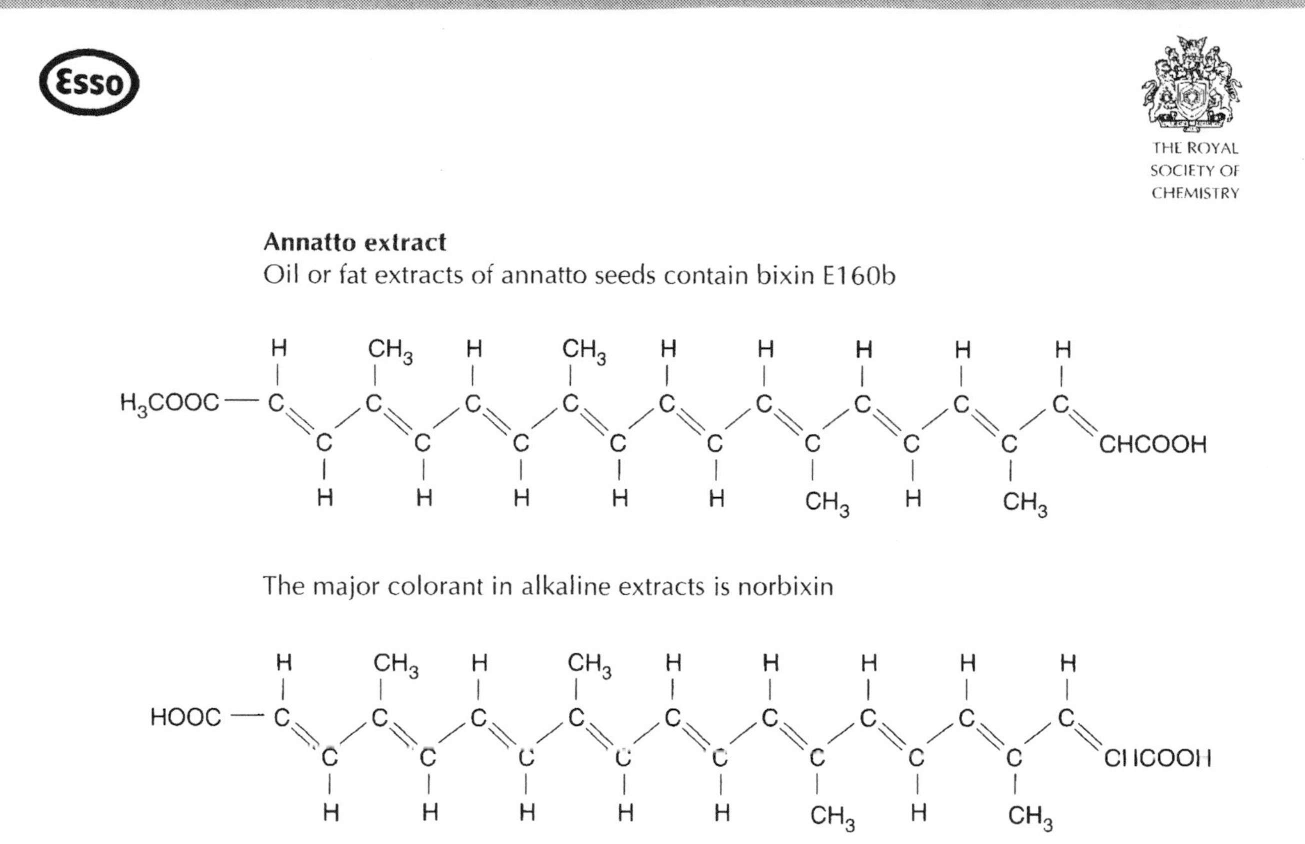

Annatto extract

Oil or fat extracts of annatto seeds contain bixin E160b

The major colorant in alkaline extracts is norbixin

Betanin E162

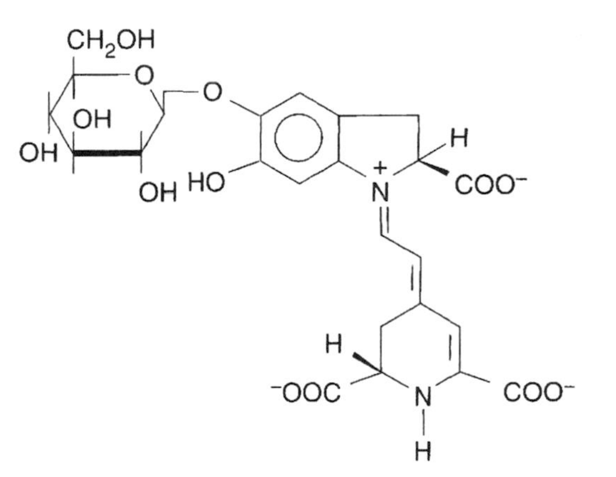

Curcumin E100

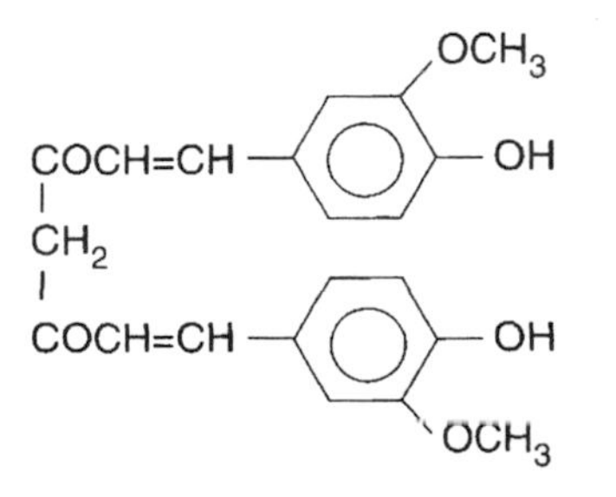

12. A transient red colour: the aqueous chemistry between iron(III) ions and sulphur oxoanions

Time

1–1.5 h.

Level

A-level, Higher Grade or equivalent.

Curriculum links

Sulphur chemistry – much of this can be looked up in text books by students. Bonding in sulphur oxoanions. Redox reactions.

Group size

2– 4.

Materials and equipment

Materials per group

- 20 cm^3 of 1 mol dm^{-3} solutions of each of:
  sodium sulphide (Na_2S)
  sodium thiocyanate (NaSCN)
  sodium sulphate (Na_2SO_4)
  sodium sulphite (Na_2SO_3)
  sodium thiosulphate ($Na_2S_2O_3$)
  sodium metabisulphite ($Na_2S_2O_5$)
  sodium dithionite ($Na_2S_2O_4$)
  sodium pyrosulphate ($Na_2S_2O_7$)
  sodium tetrathionate ($Na_2S_2O_6$)
  sodium dithionate ($Na_2S_4O_6$)
  sodium persulphate ($Na_2S_2O_8$).

Equipment per group

- test-tubes, test-tube racks
- spatulas
- dropping pipettes
- aqueous 0.5% (w/v) iron(III) chloride solution
- safety glasses.

Safety

Eye protection must be worn.

Risk assessment

A risk assessment must be carried out for this activity.

Commentary

Students should be encouraged to draw out the structures of the sulphur-based anions and consider how the anions might interact with the Fe^{3+} ion.

If 3 cm^3 of the sulphur anion solution is mixed with the same volume of iron(III) solution then the solutions in bold type will give a transient red colour:

sodium sulphide (Na_2S)

sodium thiocyanate (NaSCN)

sodium sulphate (Na_2SO_4)

sodium sulphite (Na_2SO_3)

sodium thiosulphate ($Na_2S_2O_3$)

sodium metabisulphite ($Na_2S_2O_5$)

sodium dithionite ($Na_2S_2O_4$)

sodium pyrosulphate ($Na_2S_2O_7$)

sodium tetrathionate ($Na_2S_2O_6$)

sodium dithionate ($Na_2S_4O_6$)

sodium persulphate ($Na_2S_2O_8$).

Possible answer

The correct answer is not known. A connection between the observations[1] is that those sulphur ions that can bond to iron through sulphur give the red colour – *eg*

$$\left[\mathrm{Fe}-\mathrm{S}\left(\mathrm{O}\right)_2-\mathrm{O}\right]^+$$

This is a possible intermediate in the redox reaction.

Reference

1. For further background see N. V. Reed, *School Sci. Rev.*, 1986, 768.

13. An analysis of coloured sands

Time

This activity lends itself to an extended investigation.

Once the method of analysis has been developed this could be used as a 2 h class experiment with another group.

Level

A-level, Higher Grade or equivalent.

Curriculum links

Inorganic chemistry of iron compounds. Access to suitable textbooks may be required.

Group size

1– 4.

Materials and equipment

Materials per group

- 5 g of each sample of sand (samples of sand are likely to be available locally. Packets of different colours are available by post from The Needles Pleasure Park, Alum Bay, Isle of Wight PO39 0JD. Tel: 01983 752401)

The reagents required will vary according to the method of analysis selected; those mentioned in the suggestions below are all cheap and widely available. They include:

- 10 g ammonium chloride
- 50 cm^3 of 0.5 mol dm^{-3} sodium hydroxide
- conc hydrochloric acid
- 10 cm^3 of 2 mol dm^{-3} hydrochloric acid
- 10 cm^3 of 8 mol dm^{-3} sulphuric acid
- deionised water
- 1 dm^3 standard iron(III) solution (0.1 g dm^3 or 100 ppm). This solution is prepared by dissolving 0.863 g ammonium iron(III) sulphate ($NH_4FeSO_4.12H_2O$) in 100 cm^3 of conc hydrochloric acid. The solution is diluted and transferred to a 1 dm^3 volumetric flask. It is then diluted up to the mark with deionised water and mixed well.
- potassium thiocyanate solution. This solution is prepared by dissolving 29.1 g potassium thiocyanate (KSCN) in 100 cm^3 deionised water.

Equipment per group

For the cold finger method:

- boiling tube
- small test-tube
- Bunsen burner

Esso

- safety glasses
- retort stand and clamps.

For the colorimetric method:

- colorimeter with green filter (λ_{max} 490 nm), optically matched cuvettes or colorimeter tubes
- four 10 cm^3 graduated pipettes
- eight 100 cm^3 volumetric flasks.

For the titration

- 50 cm^3 burette
- assorted pipettes
- white tile
- conical flasks
- glass rod.

Safety

Eye protection must be worn.

Risk assessment

A risk assessment must be carried out for this activity.

Commentary

The origin of the natural colours in coastal landscapes is interesting. The students are invited to apply classical chemical techniques to the problem of analysing sand. The experiments could feature as part of a special study of geochemistry.

The problem could be set in a vocational context by relating it to the manufacture of glass.[1] The sand used in glass manufacture is taken from glass sand quarries. Even a trace of iron will give a greenish colour to the glass. Sand with a low iron(III) oxide content (less than 0.25%) is therefore used in the glass industry. To achieve this the sand is treated with hot sulphuric acid to remove the iron(III) oxide. Hydrochloric acid reacts more readily but it is too corrosive to the equipment. Sulphuric acid is also available as waste from other processes hence it is cheaper. The iron(III) oxide content must not exceed 0.03% if the glassware produced is to be colourless.

In the sublimation process, iron(III) chloride forms because the ammonium chloride dissociates, forming hydrogen chloride which reacts with the iron(III) oxide in the sand. The iron(III) chloride sublimes and condenses on the cold finger with the ammonium chloride. When the sublimate is dissolved the solution contains ammonium chloride which may interfere with subsequent analyses. The iron(III) can be precipitated as iron(III) hydroxide by adding sodium hydroxide solution. The mixture can then be filtered and the precipitate redissolved in acid.

Possible approaches

(i) Colorimetric method

The red colour of the iron(III) thiocyanate complex can be used to determine the iron(III) ion concentration by a colorimetric method. The differences in the colours of the samples can be detected by eye, but it is also possible to make more precise measurements. A detailed account of a similar experiment has been given by John

THE ROYAL SOCIETY OF CHEMISTRY

Sleigh who uses this method to determine the amount of iron(III) in a metre length of audio tape.[2] A drawback to using the thiocyanate reaction is the fading of the colour with time, but this is not a problem with the procedure that he describes which is simple and rapid. The experiment is carried out in an acid medium which is an advantage because this suppresses the hydrolysis of the iron(III) thiocyanate complex.

The calibration curve for the colorimeter may be constructed[3]. Solutions containing suitable concentrations of iron(III) are prepared by diluting the stock solution (see below). At each concentration a large excess of the thiocyanate is added to transform all the iron(III) salt into the thiocyanate and so produce the maximum coloration.

Preparation of stock iron(III) solutions

Aliquots of 1.0, 1.5, 2.0, 2.5, 3.0 and 3.5 cm^3 of stock iron(III) solution (0.1 g dm^3) are pipetted into separate 100 cm^3 volumetric flasks. Each solution is diluted to about 50 cm^3 with deionised water, then 10 cm^3 of the potassium thiocyanate and 10 cm^3 of concentrated hydrochloric acid are added. The solutions are then diluted to the mark with deionised water and mixed well. (The solutions contain 1.0, 1.5, 2.0, 2.5, 3.0 and 3.5 ppm of iron(III) respectively.)

The absorbances of these standard solutions should be measured immediately, with a sample containing the same amount of potassium thiocyanate and hydrochloric acid in the reference cell. A calibration curve can be constructed showing absorbance against ppm of iron(III). As the coloration of the iron(III) thiocyanate is sensitive to the presence of other ions it is important that the sample solution should be diluted and treated in a similar way.

Preparation of sample solution

A known amount of the sample solution is transferred to a 100 cm^3 flask. It is diluted to about 50 cm^3 with deionised water. 10 cm^3 of the potassium thiocyanate solution and 10 cm^3 of concentrated hydrochloric acid are added and the solution is made up to the mark with deionised water. The absorbance is measured immediately and the concentration read off from the calibration curve.

(ii) Titration

It is also possible to determine the iron(III) concentration by volumetric methods. The amounts of iron(III) are small and the sensitivity of the various alternatives must be taken into account.

(a) The iron(III) may be titrated against EDTA. In a method given in an A-level practical textbook[3] a 6% w/v solution of sodium 2-hyroxybenzoate (salicylic acid) in propanone may be used as an indicator. The indicator recommended in Vogel's *Quantitative chemical analysis*[4] is variamine blue. The solutions are buffered to pH 2–3.

(b) Alternatively the iron(III) may be reduced to iron(II), by using zinc, and then titrated against potassium manganate(VII).

Extension

Metal ions may be determined by using atomic absorption spectroscopy. The results from the analysis may be compared with those from this instrumental technique.

Iron(III) oxides are used as pigments in cosmetics. The iron(III) oxide content of two shades of face powder could be compared by methods similar to those used with sand.

References

1. E. Barret and L. Beskeen, *Let's look at sand* produced by the Mineral Industry Manpower and Careers Unit.
2. J. Sleigh, *School Sci. Rev.*, 1988, **70,** 251.
3. B. Ratcliff, *A-Level practical chemistry*. Cambridge: CUP, 1990.
4. G. H. Jeffery et al, Vogel's *Textbook of quantitative chemical analysis*, 5th edn. Harlow: Longman. 1989.

Acknowledgements

This activity is based on suggestions from Colin Johnson, Eileen Barret and Richard Blair-Gould. Bob Mudd tested the experimental methods described above.

14. What is honey made of? The optical rotation of natural sugars

Time

2 h.

Level

A-level, Higher Grade or equivalent.

Curriculum links

Carbohydrates, chirality, optical rotation.

Group size

2–3.

Materials and equipment

Materials per group

- 30 g honey.

Equipment per group

- polarimeter. A simple polarimeter may be built[1] as part of the activity (see Activity 1)
- safety glasses.

Safety

Eye protection must be worn.

Risk assessment

A risk assessment must be carried out for this activity.

Commentary

In discussing the chirality of naturally occurring molecules students will learn that living systems are stereospecific.[2] Proteins are built almost exclusively from chiral amino acids (the L isomers) and carbohydrates are based on D sugar molecules. Honey, a natural substance, contains sugars that are comprised of D molecules.

Honeys made from the nectar of flowers are laevorotatory, a finding which should bring home the idea that D molecules are associated with the laevorotatory isomer.

Honey is a very complex substance in terms of the number and complexity of its constituents[3] but the largest proportion of the dry matter present consists of sugars. It is laevorotatory because fructose, which has a negative specific rotation, is present in the greatest quantity. The specific rotations are[4]:

sucrose	$[\alpha]_D^{20}$	+66.5°
glucose	$[\alpha]_D^{20}$	+52.5°
fructose	$[\alpha]_D^{20}$	–92.5°

Glucose and fructose together make up about 70% of the total; disaccharides including sucrose add about 10%. Only 17–20% of honey is water.

If the honey is dextrorotatory it is either honeydew honey or it has been adulterated. This was a useful test in the days when unscrupulous suppliers were likely to add cane sugar or corn syrup to honey. Today the result does not necessarily indicate purity as the honey may have been adulterated with high fructose syrup. Honeydew is produced by plant-sucking insects feeding on natural exudations of plants and it is gathered by bees which convert it into a type of honey that contains less fructose.

Procedure

A concentration of 26 g of honey to 100 cm^3 water is recommended in AOAC Official Methods of Analysis (1990).

Immediately after diluting the honey with water the optical rotation will change by a few degrees over several hours. A change of 3.5° in 20 h has been quoted.[4] This phenomenon is known as mutarotation but for honey the mechanism is not fully understood.

Some types of honey in solution are sufficiently opaque to cause problems because of light scattering and absorption during polarimetry. In such cases the honey solution can be clarified by heating rapidly to boiling, then filtering. The solution must be used immediately.

Extension

It is possible to separate and identify the sugars in honey by paper chromatography.[5]

The plant sources of honey can be identified by pollen analysis; pollen grains from different plants can be distinguished under the microscope.[6]

Other physical properties of honey are of interest. The water content of honey may be estimated from refractive index measurements. A few honeys are known for their non-Newtonian flow properties, for instance ling heather honey is thixotropic; this is due to the relatively high concentrations of certain proteins in this honey.[6]

Information about bees and honey may be obtained from the International Bee Research Association (IBRA), 18 North Road, Cardiff, CF1 3DY. An excellent book on honey, written by Eva Crane[6], is available from IBRA.

References

1. *Revised Nuffield advanced science chemistry teachers guide II.* Harlow: Longmans, 1988.
2. P. W. Atkins, *Molecules.* New York: Scientific American Library, 1987.
3. *Honey: a comprehensive survey,* E. Crane (ed). London: Heineman, 1975.
4. C. A. Browne and F. W. Zerban, *Physical and chemical methods of sugar analysis.* Chichester: J. Wiley, 1941.
5. J. N. Adds, *School Sci. Rev.*, 1992, **74**, 94.
6. E. Crane, *A book of honey.* Oxford: OUP, 1980.

15. A yellow solid

Time

The preparation requires 2 h but during this time the reaction mixture is left to stand for 1 h.

The qualitative investigation requires 1 h.

Level

A-level, Higher Grade or equivalent.

Curriculum links

Chemistry of copper(I) and copper(II) ions.

Group size

2.

Materials and equipment

Materials per group

- 10 g copper(II) sulphate
- 18 g sodium thiosulphate
- deionised water
- propanone
- 0.1 mol dm^{-3} potassium dichromate(VI) solution
- dilute nitric acid
- iodine solution (0.2 mol dm^{-3} iodine in aqueous potassium iodide)
- dilute ammonia solution
- ammonium persulphate.

Equipment per group

- two 100 cm^3 beakers
- 250 cm^3 beaker or flask
- magnetic stirrer or glass rod
- filter pump
- Buchner funnel and flask
- arrangement for heating solutions to about 40 °C
- test-tubes, test-tube rack
- filter paper
- safety glasses.

Safety

Eye protection must be worn.

Risk assessment

A risk assessment must be carried out for this activity.

Commentary

The bright yellow solid is the complex salt sodium copper(I) thiosulphate. W.G. Palmer [1,2] describes a quantitative investigation of this compound. The results quoted agree with the formula $3Cu_2S_2O_3.2Na_2S_2O_3.6H_2O$.

The results of the investigations are as follows:

(1) Sulphur dioxide gas is produced. The filter paper soaked in acidified potassium manganate(VII) turns green.

$Na_2Cr_2O_7(aq) + 3H_2SO_3(aq) + H_2SO_4(aq) \rightarrow Na_2SO_4(aq) + Cr_2(SO_4)_3(aq) + 4H_2O(l)$

(2) Sulphur dioxide is again produced.

$Na_2S_2O_3(aq) + 2HNO_3(aq) \rightarrow SO_2(g) + H_2O(l) + S(s) + 2NaNO_3(aq)$

(3) Iodine solution is decolorised and a colourless solution of sodium tetrathionate is formed.

$2Na_2S_2O_3(aq) + I_2(aq) \rightarrow Na_2S_4O_6(aq) + 2NaI(aq)$

(4) A blue copper-based complex is formed.

(5) Oxygen gas is released, indicating displacement of $S_2O_3^{2-}(aq)$ from the copper complex.

Extension

The quantitative analysis described by W.G. Palmer could be carried out.

Reference

1. W. G Palmer, *Experimental inorganic chemistry*. Cambridge: CUP, 1965.

2. R. S. Vowles, N. Bona, *School Sci Rev.*, 1985, **66**, 476.

Acknowledgement

This activity is based on a suggestion by Colin Johnson.

16. Detecting diluted evidence: screening body fluids for drugs of abuse

Time

This problem is suitable for project work over a few weeks.

Level

A-level, Higher Grade or equivalent.

Curriculum links

Organic separation techniques.

Group size

1–3.

Materials and equipment

Materials per group

- 1 foil-wrapped tablet of ASPRO-CLEAR
- 2 cm^3 of 4 mol dm^{-3} hydrochloric acid
- 2 cm^3 Trinder's reagent (see below)
- 100 cm^3 'horse urine' sample (see below)
- 25 cm^3 methanol
- 25 cm^3 dichloromethane
- 2 g XAD-2 Resin (20/60 mesh available from: Alltech, 6–7 Kellet Road Industrial Estate, Kellet Road, Carnforth, Lancashire LA5 9XP. Tel: 01524 734451).

Preparation of Trinder's Reagent
This solution should be prepared in advance. It may be stored at room temperature and will keep for one year. To prepare it you will need:

- 15 cm^3 of 2 mol dm^{-3} hydrochloric acid
- 10 g mercury(II) chloride, $HgCl_2$
- 10 g iron(III) nitrate $Fe(NO_3)_3.9H_2O$
- deionised water.

10 g of mercury(II) chloride is dissolved in approximately 210 cm^3 of deionised water, with warming if necessary.

10 g of iron(III) nitrate and 15 cm^3 of 2 mol dm^{-3} HCl are added and the mixture is transferred to a 250 cm^3 volumetric flask. The solution is made up to volume with deionised water.

The solution should be transferred to a bottle. This should be labelled with the appropriate hazard symbol to indicate that the solution is toxic.

Preparation of 'horse urine' sample
Fill a 500 cm^3 beaker with water and colour it with a very small amount of tartrazine.

Equipment per group

- two 1 cm^3 pipettes
- three 100 cm^3 beakers
- 5 test-tubes in a test-tube rack
- anti-bumping granules
- glass rods
- glass bottle
- deionised water
- Bunsen burner, tripod and gauze
- safety glasses.

To test out practical solutions to the problem students will need various items of readily available laboratory equipment including:

- 2 cm^3 plastic disposable syringe without needle
- glass wool
- small glass funnel
- retort stand
- boss
- clamp
- ring clamp
- 100 cm^3 separating funnel.

The professional solution to the problem is to use miniature disposable solid phase extraction cartridges. They are not necessary for this problem-solving exercise but might provide added interest if they are available from a college biochemistry department. A suitable type is: C18 Bond Elut® cartridges, column volume 2.8 cm^3. The UK supplier is: Jones Chromatography, Tir-y-Berth Industrial Estate, New Road, Hengoed, Mid Glamorgan CF8 8AU. Tel: 01443 816991.

Safety

Eye protection must be worn.

A fume cupboard should be used where appropriate, especially if a solvent extraction using dichloromethane is attempted.

Risk assessment

A risk assessment must be carried out for this activity.

Commentary

Chemists in forensic laboratories provide analyses and interpret them for the police, the courts, the health and social services, and sporting bodies. Besides working with great accuracy the chemist needs to be able to think creatively as the interpretation of results may be complex.

One of the main problems facing chemists who analyse body fluids such as urine or blood is that the compounds they are interested in finding are often present at very low concentrations. What is generally needed from such tests is a 'yes' or 'no' answer without the possibility of any false positive result. With salicylic acid there is a

threshold value because horses may ingest certain plant materials which will lead to very small amounts of salicylic acid being present in the urine of animals that have not been given aspirin.

An enormous range of compounds is present in horse urine. A more detailed account of the background to this problem is given in an earlier RSC publication.[1] After a race, samples are collected by the veterinary officer's assistant. The Jockey Club will hold an enquiry if a horse fails a drug test and this may lead to the trainer being fined or, if the doping was deliberate, losing his licence. If a horse has been 'nobbled' the police are called in as this is a criminal offence.

Possible approaches

One possibility is to use solvent extraction. The salicylic acid could be extracted by using an immiscible solvent such as dichloromethane. The solvent can be evaporated in the fume cupboard. The salicylic acid is then dissolved in a small amount of methanol when it should react positively with Trinder's reagent.

Liquid/liquid extraction methods tend to be plagued by emulsion formation. In the 1970s resins were developed which could remove trace organic contaminants from water and from body fluids.[2] A resin which is used widely is XAD-2 resin, a non-ionic styrene-divinylbenzene copolymer. The aqueous sample is passed through a column of the resin: the organic substances are adsorbed and can then be eluted using methanol. This technique is known as solid phase extraction and it may be necessary to carry out a few experiments to optimise conditions.

One method is to use a disposable plastic 2 cm^3 syringe without a needle. A plug of glass wool is carefully inserted above the constriction in the syringe. 2 g of dry XAD-2 resin is made into a slurry with methanol and poured into the syringe to make a column approximately 6 cm high. The column is washed first with methanol and then with deionised water. The 50 cm^3 'urine' sample is then poured down the column. The flow rate can be adjusted by using the plunger from the syringe.

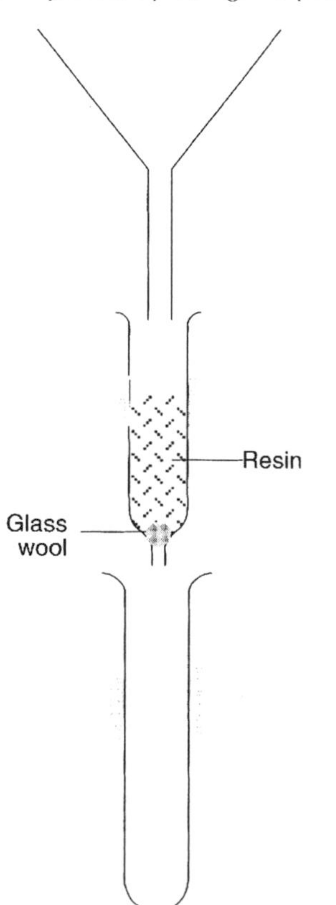

1 cm^3 of methanol is then put into the column and the plunger is inserted so that the methanol does not drain through the column. The methanol is allowed to remain in contact with the resin for about half an hour. The plunger is then removed and the methanol runs out.

This sample should give a strong purple colour with Trinder's reagent. Further amounts of methanol can be put through the column and tested to see how effective the extraction is.

Extension

Analytical laboratories have been required to develop new techniques for screening very large numbers of urine samples. Special disposable solid phase extraction columns have been developed.

The principles on which they are based are similar to those used in the XAD-2 column above and these contain chemically modified silica gels.

If these cartridges are available it is interesting to see how effective they are, but they are not essential for this exercise. In an analytical laboratory the cartridges are components in an automated system but it is easy to show how well they work with a makeshift arrangement. The 50 cm^3 'urine' sample can be pulled through the cartridge by connecting it to a large syringe. The cartridge is then sucked dry. 1 cm^3 methanol is pulled through and tested with Trinder's reagent. If further samples of methanol are drawn through and tested this will show that almost all the salicylic acid is eluted in the first cm^3.

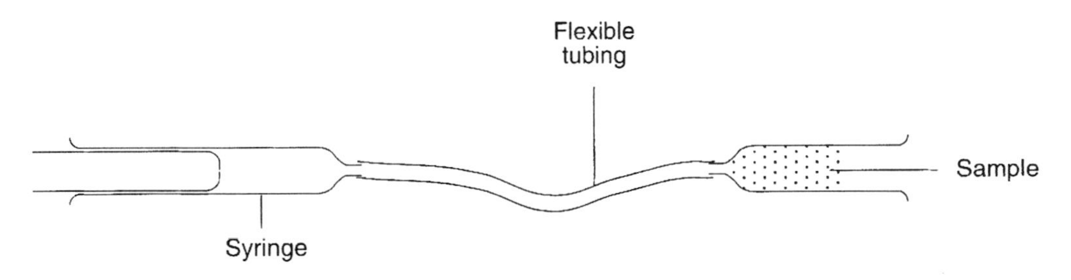

References

1. B. Faust, *Modern chemical techniques*. London: RSC, 1992.

2. G. A. Junk *et al*, *J. Chromat.*, 1974, **99,** 745

Acknowledgement

This activity is based on an idea from Mrs Pat Chalmers.

17. As sweet as? Detecting aspartame in a table-top sweetener

Time

Stage 1 Hydrolysis of Canderel®: 1–1.5 h (includes 0.5–1 h refluxing).
Stage 2 Paper chromatography: 2.5 h (includes 1–2 h for running chromatogram).

Level

A-level, Higher Grade or equivalent.

Curriculum links

Amino acids, peptide linkage. Paper chromatography.

Group size

Stage 1 Hydrolysis of Canderel®: If time is limited one batch could be hydrolysed to provide material for all the groups to use in preparing chromatograms.
Stage 2 Paper chromatography: 2–3.

Materials and equipment

Materials per group

Stage 1 Hydrolysis of Canderel®

- 12 g Canderel®
- 200 cm^3 of 6 mol dm^{-3} hydrochloric acid.

Stage 2 Paper chromatography

- about 5 cm^3 of the solution produced in Stage 1 (acid hydrolysis of Canderel®)
- 100 mg activated charcoal
- about 50 cm^3 of solvent mixture for chromatography (ethanol:water:880 ammonia in the ratio 80:10:10)
- ninhydrin, 0.2% solution in propanone, stored in a spray bottle (Ninhydrin is also available from BDH in a spray can as a 0.5% solution in butanol)

Reference amino acids

- 1 cm^3 of a DL-aspartic acid 0.01 mol dm^{-3} solution dissolved in 10% v/v propan-2-ol/water
- 1 cm^3 of a DL-phenylalanine 0.01 mol dm^{-3} solution dissolved in 10% v/v propan-2-ol/water.

Equipment per group

Stage 1 Hydrolysis of Canderel®

- 500 cm^3 round-bottomed flask
- condenser
- Bunsen burner or heating mantle
- safety glasses.

Stage 2 Paper chromatography

- pasteur pipette
- 5 cm^3 measuring cylinder
- test-tubes
- small funnel and filter paper
- chromatography tank *or* 1 dm^3 beaker and cling film to cover
- chromatography paper (Whatman No. 1)
- 25 or 50 cm^3 measuring cylinder depending on size of tank
- clips for paper, pencil

Access to:

- fume cupboard
- oven at 110 °C
- spray bottle containing 0.2% ninhydrin solution in propanone.

Safety

Eye protection must be worn.

The ninhydrin spray should be used only in a fume cupboard. The chromatogram must be hung up inside the fume cupboard to be sprayed.

Risk assessment

A risk assessment must be carried out for this activity.

Commentary

The idea of applying the techniques of chromatography to the analysis of Canderel® is based on an experiment described by A. D. Heaton.[1] If the students follow the approach suggested below they should obtain clear results as only two amino acids are involved.

Aspartame

The sweet taste of aspartame was discovered accidentally in 1965 by James Schatter who was synthesising a product for treating ulcers.[2] He was heating aspartame in a flask with methanol when the mixture bumped onto the outside of the flask. He later detected a strong sweet taste on his fingers which he traced back to powdered aspartame on the flask. This method of discovery is not an example of good laboratory practice!

Aspartame is valued because it has a clean sweet taste similar to that of sucrose. The aspartame in Canderel® is bulked out with carbohydrate so that one teaspoonful is perceived to be as sweet as one teaspoonful of sugar.

Aspartame is one of the most thoroughly tested food additives.[3] Aspartate, phenylalanine and methanol are produced when it is metabolised. The safety of aspartame has been called into question because high blood levels of each of these compounds is associated with toxicity. Because aspartame is approximately 200 times sweeter than sugar very little is needed to provide the equivalent sweetness. The amounts of the amino acids and methanol provided by a normal diet are much larger than those likely to be ingested as aspartame. A teaspoonful of Canderel® contains 20 mg phenylalanine while an 8 oz glass of milk provides 542 mg.

An infant suffering from the genetic disease phenylketonurea(PKU) is likely to be

Esso

on a phenylalanine-restricted diet as this can prevent the onset of mental retardation that is associated with the disease. In such cases aspartame should be avoided.

Procedure

The procedure for paper chromatography is adapted from a manual by Smith and Feinberg.[4] An experiment on the chromatographic analysis of amino acids forms part of the Nuffield chemistry course.[5]

Stage 1 Hydrolysis of Canderel®

12 g of Canderel® is placed in a round-bottomed flask and 200 cm^3 of hydrochloric acid (6 mol dm^{-3}) is added. The mixture is then refluxed for ½–1 h. After a short time the mixture will begin to turn brown, by the time this stage is finished it will be black.

Stage 2 Paper chromatography

Care should be taken to touch the chromatography sheets only at the top corners as fingerprints contain traces of amino acids.

A small sample of the black mixture is first decolorised by using activated charcoal. A pasteur pipette is used to transfer *ca* 5 cm^3 of the hydrolysate to a clean test-tube. This is decolorised with *ca* 100 mg activated charcoal and filtered to give a clear solution for spotting onto the chromatogram.

The solvent mixture (ethanol:water:880 ammonia) is placed in the tank which is covered to produce a saturated atmosphere.

The paper is prepared and spots of each of the reference amino acids and also the sample are placed on the paper. Pencil identification marks are made at the top of the paper.

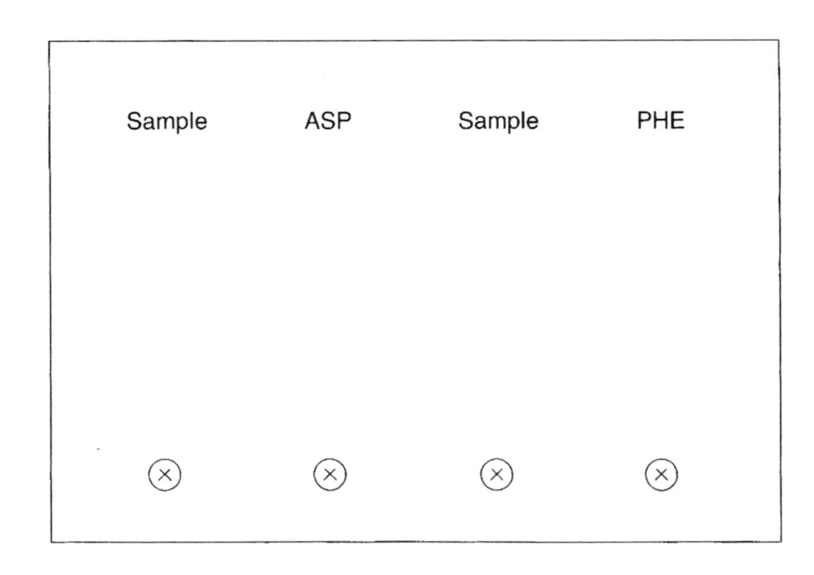

The paper is formed into a cylinder and secured with clips. It is then placed, with the spotted end down, in the tank taking care not to let the paper touch the glass walls. The tank is closed. No observations can be made while the chromatogram is running because the compounds used are colourless. The chromatogram is run for a minimum of 1 h and longer if possible. It is then removed from the tank, the solvent front is marked with a pencil, and the paper is allowed to dry.

The paper is then hung up in a fume cupboard and sprayed sparingly with the ninhydrin solution. It is then heated in an oven at 110°C for 5–10 min when the amino acids should appear as purple spots .

The colour is stable for some weeks if kept in the dark and can be photocopied to give a permanent record.

Extension

Although aspartame tastes very similar to sucrose, food chemists have to take its chemical properties into account before it is included in a food product in place of sugar. Studies of the stability of aspartame in solution have shown that it is likely to be fully hydrolysed within 9 days at pH 7.4.[2,3] If the aspartame is in a food system when this happens a loss of sweetness will be perceived. This effect could be investigated by dissolving Canderel® in a buffer solution and leaving it for various lengths of time in a warm place.

References

1. A. D Heaton, *School Sci. Rev.*, 1985, **66**, 728 .
2. *Aspartame: physiology and biochemistry*, D. Lewis, Stegink and L. J. Filer (eds). Marcel Dekker, 1984.
3. *Food additive user's handbook*, J. Smith (ed). London: Blackie, 1991.
4. I. Smith and J. G. Feinberg, *Paper and thin layer chromatography and electrophoresis*. London: Longman, 1972.
5. *Nuffield advanced science chemistry students' book II*. London: Longman, 1984.

18. Liquid and solid water; the growth of ice crystals

Time

1–2 h (depending on approach).

Level

A-level, Higher Grade or equivalent.

Curriculum links

Structure and bonding, hydrogen bonds.

Group size

1– 4 (depending on approach).

Materials and equipment

Materials per group

- ice
- water.

Equipment per group

- molecular models
- laboratory glassware – *eg* beakers
- measuring cylinders
- thermometer
- stop-watch
- safety glasses.

Safety

Eye protection must be worn if practical work is undertaken.

Risk assessment

A risk assessment must be carried out for this activity if practical work is undertaken.

Commentary

Liquid water is made up of clusters of water molecules joined together by hydrogen bonds which are continually breaking and reforming. As they turn into solid ice they become fixed into the three dimensional pattern of the crystal. The hydrogen bonds hold the molecules further apart in ice so that it is less dense than water.

This activity is intended to encourage students to think more deeply about the process of freezing. After learning about hydrogen bonding the students may explore the arrangement of the atoms in space by building a model of an ice structure. They can work in groups to do this, each student constructing a few molecules and then joining them together. Orbital molecular models are very suitable. This in itself is an exercise in group problem solving and may be done as a competition.

Possible answers

(i) The students should be able to arrive at an approximate value from their own experience. This should be about 3×10^{-3} mm s^{-1}. It is also possible to arrive at an answer by calculation:

Mathematical treatment of the freezing of a pond

There are two main methods of calculation, both based on the simple model:

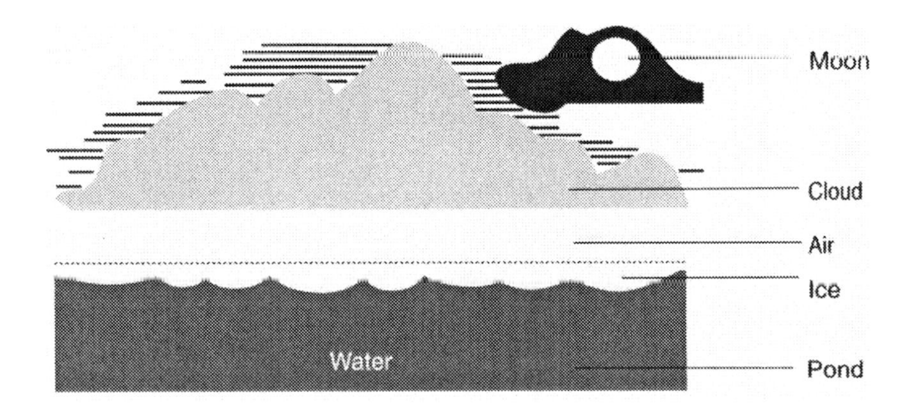

Let the air temperature above the ice remain steady at θ°C. The temperature at the interface between the ice and the water is assumed to be 0°C. The latent heat (L) is conducted away upwards.

(1) This method uses latent heat and thermal conductivity to calculate the heat losses.

density of ice ρ kg m^{-3} is given by

$$\rho = \frac{m}{V}$$

where m is mass in g, V is volume in m^3

$\rho = \frac{m}{Ax}$ where A is area in m^2, and x is thickness in metres

$$m = A\rho x$$

Quantity of heat released when a given mass m of liquid condenses is given by mL, where L is the latent heat of fusion at 0°C

So quantity of heat released = mL = AρxL

hence if the ice thickens at a rate of $\frac{dx}{dt}$ then loss of heat = $A\rho L.\frac{dx}{dt}$

However, rate of loss of heat is also given by k multplied by the temperature gradient, where k is the thermal conductivity in J s^{-1} m^{-1} K^{-1}

Let the temperature be θ, so that

rate of loss of heat = $\frac{k(\theta-0)}{x} = \frac{k\theta}{x}$

Therefore $A\rho L\frac{dx}{dt} = \frac{k\theta}{x}$

$$\frac{dx}{dt} = \frac{k\theta}{xA\rho L}$$

Let A be unit area, 1 m^2,

rate of increase in thickness of ice = $\frac{k\theta}{x\rho L}$

Data
$k = 2.1\ J\ s^{-1}\ m^{-1}\ K^{-1}$
$L = 333\ kJ\ kg^{-1}$
$\rho = 0.92 \times 10^3\ kg\ m^{-3}$
$x = 1.0 \times 10^{-2}\ m$

If it is assumed that the air temperature above the ice is –5°C, then the rate of freezing is given by

$$\frac{2.1 \times 5}{333 \times 10^3 \times 0.92 \times 10^3 \times 1.0 \times 10^{-2}}\ m\ s^{-1}$$

The ice is therefore increasing in thickness at a rate of 3×10^{-3} mm s^{-1}.

(2) This more rigorous method reaches the same final equation.

The rate at which the ice thickens is limited by the thermal conductivity (k) of the ice which has already formed.

Let ρ kg m^{-3} be the density of the ice at 0°C and k $Js^{-1}\ m^{-1}\ K^{-1}$ its thermal conductivity.

At any instant t seconds from the start of freezing, let the thickness of the ice be x metres and let the thickness increase by a small amount δx in a small further instant of time δt.

Consider unit area of the surface.

Rate of flow of heat = k multiplied by the temperature gradient

In time δt the volume of ice formed will be x m^3 (since we are considering unit area of surface). The mass of ice formed will therefore be $\rho\delta x$ kg and the quantity of heat flowing through the ice will be $L\rho\delta x$ joules, where L is latent heat of fusion at 0°C (J kg^{-1}). So, the rate of flow of heat through the ice is given by $\frac{L\rho\delta x}{\delta t}$ Js^{-1}.

The lower surface of the ice is at 0°C and the upper at $-\theta$°C, so the temperature gradient is $\frac{\theta}{x}$.

It follows from the relationship above that

$$\frac{L\rho\delta x}{\delta t} = \frac{k\theta}{x}$$

In the limit,

$$\frac{dx}{dt} = \frac{k\theta}{L\rho x}$$

Esso

The rate of thickening of the ice, $\frac{dx}{dt}$, is therefore directly proportional to θ and indirectly proportional to the existing thickness x.

$$\text{Rate of increase of thickness of ice} = \frac{k\theta}{L\rho x}$$

where ρ is the density of ice, t is the time from the start of freezing and x is the thickness.

Using the same data as before, the rate of freezing is given by

$$\frac{2.1 \times 5}{333 \times 10^3 \times 0.92 \times 10^3 \times 1.0 \times 10^{-2}} \text{ ms}^{-1}$$

The ice is therefore increasing in thickness at a rate of 3×10^{-3} mm s^{-1}.

(ii) Students could design an experiment to investigate the rate of freezing (or melting). A very simple approach would be to immerse a large block of ice in water and to observe the rate at which it melts. Alternatively, ice could be placed in a funnel.

A more accurate method, which is described in the literature, is to observe the formation of ice crystals in thin (*ca* 1 mm internal diameter) polythene tubes. The tubes are filled with water and placed in a bath at a temperature below 0°C. Freezing of the supercooled water in the tube is initiated at one end and the growth velocity of the ice along the tube is measured with a stop watch. Under these conditions the latent heat can be dissipated into the surrounding liquid. This takes place more rapidly than conductive transfer through solid ice.

(iii) At 0°C, density of ice = 0.92×10^3 kg m^{-3}

density of water = 1.00×10^3 kg m^{-3}

The structure of ice is open. Building a three dimensional model is a very effective way of seeing that a decrease in density is likely to occur when water freezes.

(iv) The answer to **(i)** applies to the formation of ice on the surface of a pond. Ice crystals can grow in supercooled liquid water or by sublimation from the vapour phase. The rates may be more than one million times greater than the value above.

The mechanisms that control the rate of growth are:

(a) transport of the molecules to the point of growth;

(b) binding of the molecules into the crystal; and

(c) removal of the latent heat from the interface.

In the case of a layer of ice on a pond the last process is very slow and is the rate-determining step. The rate at which the latent heat is conducted away slows down as the ice thickens. It can be shown that it is inversely proportional to the thickness (see **(i)**). When the ice has reached a certain thickness the rate will become effectively zero so that the pond will never freeze solid.

Extension

This topic lends itself well to a discussion based on 'What if?'[2] Students could discuss what would happen if ice were not less dense than water, if it was a good conductor of heat *etc.*

When ice grows in a living system the process causes violent changes within the cells. Cryobiologists look at the biological effects of this phase change. They study the effects of low temperature on living things – food preservation is an important application.[3]

The phase change that occurs when a liquid metal solidifies is of great importance in metallurgy.[4] Again heat dissipation is significant as it can determine the direction of growth of crystals in the melt. By controlling the heat flow it is possible to produce a turbine blade made of a single crystal.

References

1. N. H. Fletcher, *The chemical physics of ice*. Cambridge: CUP, 1970.

2. *SATIS 16–19: Science and technology in society*, A. Hunt (ed). Hatfield: ASE, 1990.

3. *Nuffield advanced science chemistry – food science, A special study*, (revised edn). London: Longman, 1984.

4. *Nuffield advanced science chemistry – metals as materials, A special study*, (revised edn). London: Longman, 1984.

Acknowledgement

This activity was developed with the help of Colin Osborne.

19. Vintage titrations: sulphur dioxide in wine

Time

1–1.5 h.

Level

A-level, Higher Grade or equivalent.

Curriculum links

Redox titrations using iodine.

Group size

2.

Materials and equipment

Materials per group

- 120 cm^3 of white wine (see below)
- 50 cm^3 of 0.01 mol dm^{-3} iodine solution (stabilized with potassium iodide)
- 20 cm^3 of 2.5 mol dm^{-3} H_2SO_4
- 25 cm^3 of 1 mol dm^{-3} sodium hydroxide
- 7 cm^3 of 2% starch solution
- deionised water.

Equipment per group

- 50 cm^3 burette
- 25 cm^3 pipette
- 10 cm^3 and 25 cm^3 measuring cylinders
- 250 cm^3 conical flasks
- white tile
- safety glasses.

Safety

Eye protection must be worn.

Risk assessment

A risk assessment must be carried out for this activity.

Commentary

This analysis is based on the familiar titration with iodine, using starch as an indicator.[1] In trialling some students needed help to devise an appropriate method; others coped easily.

Procedure

Free SO_2

50 cm^3 of wine is pipetted into a 250 cm^3 conical flask and *ca* 5 cm^3 of sulphuric acid and 2–3 cm^3 of starch solution added.

The solution is titrated with 0.01 mol dm^3 I_2 solution. The end-point is taken to be the appearance of a dark blue colour which persists for about 2 minutes.

In the interests of economy one very careful titration should be sufficient.

Total (free and combined) SO_2

25 cm^3 of 1 mol dm^{-3} NaOH is placed in a 250 cm^3 conical flask, using a measuring cylinder. 50 cm^3 of wine is pipetted into this flask. The flask is shaken and left to stand for 15 minutes then 10 cm^3 of sulphuric acid and 2–3 cm^3 of starch solution are added. The solution is titrated with 0.01 mol dm^3 I_2 solution as above.

Calculation

The amounts of free and combined SO_2 can be calculated as mol dm^{-3} and as mg dm^{-3} (parts per million or ppm). The legal limit for total SO_2 varies from one country to another; 250 ppm is a commonly accepted value.

Although there is no legal limit on the amount of free SO_2, levels from 20–40 ppm safeguard the wine without affecting its taste. If the level is below 10 ppm in a white wine it is in danger of going bad.

Extension

The method is not usually recommended for red wines because the colour masks the end-point. However it can normally be seen without too much difficulty if the mixture in the flask is compared with a sample of the original wine.

Reference

1. G. F. W. Fowles, *Educ. Chem.*, 1978, **15**, 89.

(Esso)

20. Vintage titrations: tannin in wine

Time

2 h.

Level

A-level, Higher Grade or equivalent.

Curriculum links

Redox titrations using potassium manganate(VII).

Group size

2.

Materials and equipment

Materials per group

- 50 cm^3 samples of red wine (for white wine see below)
- 50 cm^3 of 0.004 mol dm^{-3} potassium manganate(VII)
- 1 g activated charcoal
- deionised water
- 10 cm^3 of 0.5% indigo carmine indicator solution. (The indigo carmine indicator is made up by dissolving 0.5 g of the dyestuff in 60 cm^3 warm deionised water. The solution is cooled, 4 cm^3 of conc H_2SO_4 is added, and the volume made up to 100 cm^3 with deionised water. The solution is filtered through a No. 42 Whatman paper).

Equipment per group

- 50 cm^3 burette
- 5 cm^3 and 2 cm^3 pipettes
- funnel
- filter paper
- 10 cm^3, 25 cm^3 and 250 cm^3 measuring cylinders
- 250 cm^3 conical flasks
- 50 cm^3 beaker
- white tile
- safety glasses.

Safety

Eye protection must be worn.

Risk assessment

A risk assessment must be carried out for this activity.

Commentary

The colour of red wine is due to the presence of anthocyanidins, a class of flavonoids.[1] Tannin is a collective name for other largely colourless but bitter flavonoids which are also present in the wine. In making red wine the crushed grapes are put into vats. Some of the stems, the skins, and the pulp remain with the juice forming a residue which is known as the must. The alcohol in the fermenting juice extracts colour from the skins and the longer the juice is in contact with the must the darker the wine. In the process tannin is also extracted into the wine.

White wine, which should not pick up any colour from the grape skins, is made by pressing the grapes as quickly as possible and the juice alone is then set to ferment. The level of tannin in white wines is only about one-tenth of that found in red wine.

The procedure described below is based on that described by Professor G.W.A. Fowles of University of Reading.[2] During trialling most students needed a lot of guidance to work through this procedure.

Procedure

Actual titration

A funnel is placed in the neck of a 250 cm^3 conical flask. 5 cm^3 of wine is pipetted and 10 cm^3 of deionised water is added. The flask is heated gently until the volume of the wine and water is reduced to 5–7 cm^3. The alcohol will now have boiled off. 25 cm^3 of cold deionised water is now added and 2 cm^3 of indigo carmine indicator is added by using a pipette. (As the indicator uses up some potassium manganate(VII) it is important to measure it out carefully so that it can be allowed for in the 'blank'.)

0.004 mol dm^3 $KMnO_4$ is placed in the burette and the mixture is titrated. A golden yellow colour appears at the end point. Let this titre be A cm^3.

Blank titration

20–25 cm^3 of the wine is placed in a beaker with 1 g of activated charcoal and the mixture is stirred thoroughly. The mixture is then filtered and 5 cm^3 of the decolorised wine is transferred to a 250 cm^3 conical flask using a pipette.

The procedure described in *'Actual titration'* is repeated. This blank titration will allow for the indicator and for any oxidisable substances in the wine apart from the anthocyanidins and the tannins.

Let the volume of the blank titre be B cm^3.

Calculation

The amount of potassium manganate(VII) used in oxidising the tannins (and anthocyanidins) is A – B = C cm^3.

Tannins are of variable composition. The titration is referred to a standard tannin solution for which 1 cm^3 of 0.004 mol dm^{-3} $KMnO_4$ = 0.0832 mg tannin. Therefore % tannin in wine = 0.01664 C

The level of tannin for burgundies and clarets will be in the range 0.15 – 0.4%.

Extension

In his book *Chemistry in the Marketplace* Ben Selinger describes a series of experiments on wine analysis[3] with Australian wines which are designed for first year undergraduates.

References

1. P. W. Atkins, *Molecules.* New York: Scientific American Library Series, 1987.
2. G. F. W. Fowles, *Educ. Chem.*, 1978, **15,** 89.
3. B. Selinger, *Chemistry in the marketplace,* 4th edn. London: Harcourt Brace Jovanovich, 1989.

21. A taste for kilojoules: food calorimetry

Time

2 h.

Level

A-level, Higher Grade or equivalent.

Curriculum links

Chemical energetics, food chemistry.

Group size

2–3.

The number of students may be determined by the number of groups that can be comfortably supervised. This activity lends itself to team work because each measurement involves a series of different practical steps and members of the team can take it in turn to do each of these as the measurements are repeated. The assistance of a technician was found to be useful in running this activity.

Materials and equipment

Materials per group

- Supply of oxygen
- Food samples: bread (dried), cornflakes, oils, nuts, sugar, meat (dried), celery (dried), cabbage (dried), crisps, savory snacks, sweet snacks.

Equipment per group

- Food calorimeter (see *Commentary*), including: glass calorimeter vessel, perspex cover, loop stirrer, copper helix with bung, support table, igniter support and leads, nickel crucible, spare igniter coils.
- Retort stand with large clamp
- 500 cm^3 measuring cylinder
- – 5° to 50° x 0.1°C or –10° to 110°C x 1°C thermometer
- 6V power supply
- safety glasses.

Safety

Eye protection must be worn. The oxygen supply should not be used with liquid fuels – *eg* alcoholic spirits. If a liquid fuel is used it should be placed in a small spirit lamp, immediately capped with the ground glass cover to prevent evaporation, and its mass found. A filter pump and connecting tubing is used to draw air through the calorimeter to support combustion. The equipment is similar to that used in the traditional exercise to determine the heats of combustion of a homologous series of alcohols.

Risk assessment

A risk assessment must be carried out for this activity.

Commentary

While it is possible to construct or buy a calorimeter during trialling it was found that many biology departments had apparatus of this type.

It is possible to obtain data for calculating the heat capacity of the calorimeter, for which purposes the masses of the relevant components must be found. Although the apparatus is designed to transfer most of the energy of the flame to the calorimeter, a significant amount is lost to the surrounding air. The calculation of the heat taken up by the calorimeter involves approximations *eg* the outside of the apparatus is cooler than the inside. These difficulties may be overcome if the calorimeter is calibrated by using a food sample of known energy value.

Some food samples will need to be dried and simple calculations are necessary to allow for the loss of water. Weaker students who have difficulty with the concept of ratio may need help and for this reason it may be advisable to restrict them to dry foods.

Information on food energy values is often available on packaging but it may also be obtained from a variety of books.[1]

Procedure

Detailed instructions may be found in the technical manual that accompanies the food calorimeter. The procedures for some common foods are:

1. Bread (dried)

Weigh accurately between 0.25 and 0.5 g of sample into a crucible. Place the crucible on a heat resistant platform and clamp a calorimeter on top of the platform. Turn on the oxygen supply, note the water temperature and ignite the sample. Lower the filament onto the food and switch on. Once the food has ignited, remove the filament by swinging it to the side. Stir the water continuously after combustion has ceased, while regulating the oxygen supply to promote controlled burning. Record the final temperature of the water.

2. Olive oil (see *Safety*)

Take up about 0.3 cm^3 of oil from a syringe in a tuft of 'asbestos' wool by making this into a pyramid with a tapering wick at the top. Place this in a crucible on an asbestos platform.

3. Sugar

Heat one corner of a sugar lump gently in a micro-Bunsen flame until soft, then press it into a little cigarette ash (this contains compounds of lithium that catalyse combustion).

Calculation

An effective method of tabulating the results should be devised and cooling curves plotted to calculate the theoretical maximum temperature rise. During trialling some students made use of spreadsheets to find the nutritional value of their daily food intake.

The calculation is based on the energy change being related to the product of the mass of water x specific heat capacity of water x temperature rise.

Extension

The practical investigation could be extended by for example, comparing the energy values of fried chips and oven chips. Some students may like to hypothesize that a piece of celery supplies less energy than it takes to eat it, while others might consider hot versus cold foods.[2]

References

1. B. Holland *et al*, McCance and Widdowson's *The composition of foods*, 5th edn. London: RSC, 1992.

2. D. Archer, *What's your reaction?*. London: RSC, 1991.

Esso

22. Theory v practice: do they compare?

Time

2–3 h for the theoretical approach depending on ability and on amount of assistance that needs to be given.

2– 4 h for the practical approach.

Level

A-level, Higher Grade or equivalent.

Curriculum links

Born-Haber cycle calculations, hydration enthalpies, practical calorimetry.

Access to data books will be needed. The book by Stark and Wallace[1] contains the most useful compilation of data for this exercise.

Group size

Minimum of 3 for theoretical work (*ie* for group discussion).

Materials and equipment

Materials per group

- 0.5–1.0 g fresh calcium metal
- 60 mm x 60 mm copper gauze to make a cage in which to sink the calcium.

Equipment per group

- 2 dm^3 beaker (plastic if possible)
- 0°–100° x 0.1°C thermometer
- safety screen or fume cupboard
- safety glasses.

Safety

Eye protection must be worn. When attempting the practical route students must have their proposals approved before starting on any experimental work. The use of a safety screen or fume cupboard should be mandatory.

Risk assessment

A risk assessment must be carried out for this activity.

Commentary

Various versions of the cycle are possible, but if the students invoke the electron affinity of the hydroxide radical, which is given in the brief, they should eventually arrive at the following:

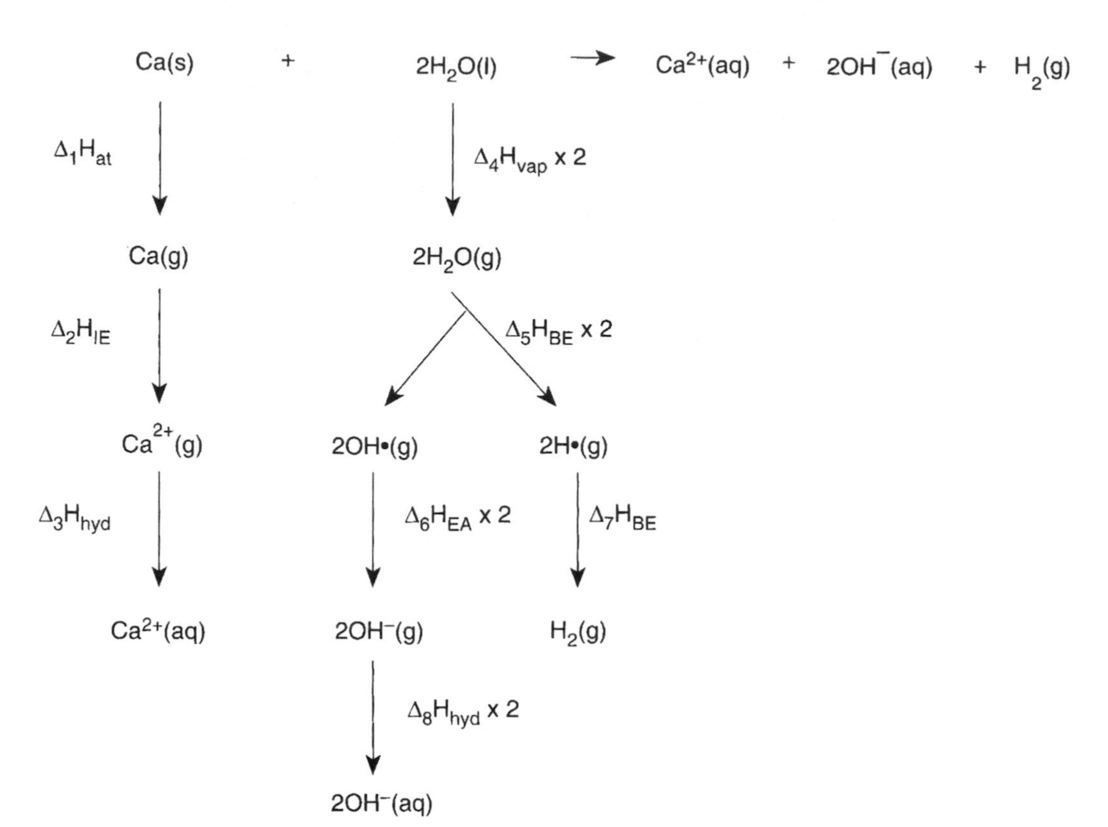

Energy values ΔH/kJ mol^{-1} (from data books and textbooks)[1]

$\Delta_1 H_{at}$ = atomisation of Ca(s) =	+ 193
$\Delta_2 H_{IE}$ = 1st and 2nd ionisation energies of Ca = +590 +1150 =	+ 1740
$\Delta_3 H_{hyd}$ = hydration energy of Ca^{2+} =	–1650
$\Delta_4 H_{vap}$ = vaporisation of water x 2 = +44 x 2=	+ 88
$\Delta_5 H_{BE}$ = bond energy O–H x 2 = +463 x 2 =	+ 926
$\Delta_6 H_{EA}$ = electron affinity of OH•(g) = –176.5 x 2=	– 353
$\Delta_7 H_{BE}$ = bond energy H–H = –436	– 436
$\Delta_8 H_{hyd}$ = hydration energy of OH^-(aq) = –460 x 2	– 920
	– 412 kJ mol^{-1}

Procedure

The experimental work is straight forward and gives results within 95% of the theoretical value using the following method.

Weigh about 100.00 g of deionised water in a 2 dm^3 beaker. Carefully wipe the inside of the beaker to remove droplets above the water. Stand the beaker in a fume cupboard, insert a thermometer capable of reading to 0.1°C in the water and leave to obtain a constant value. Weigh out about 0.5 g of calcium metal, choosing large pieces rather than smaller ones. Quickly empty the metal into the water and stir with the thermometer. If the calcium is fresh the reaction will be complete in *ca* 90 s and the temperature increase is roughly 1.2°C.

Some of the calcium floats, despite its density, because of the rapid evolution of hydrogen, hence some of the heat is lost to the air rather than to the water. To circumvent this problem during trialling some students designed a copper cage to ensure that the calcium remained submerged. If this is done a correction factor for the copper will be needed.

The accuracy of the experiment can be improved by using lagging and plotting cooling curves to find the theoretical maximum temperature rise.

Calculation

The experimental calculation is based on the energy change being related to the product of the mass of water x specific heat capacity of water x temperature rise.

Reference

1. J. G. Stark and H. G. Wallace, *Chemistry data book*. London: John Murray, 1982.

Acknowledgement

This activity is based on a problem written by Joe Burns.

23. Three isomeric alcohols

Time

1–1.5 h.

Level

A-level, Higher Grade or equivalent.

Curriculum links

Reactions of primary, secondary and tertiary alcohols. Reactions of aldehydes and ketones.

Group size

1–2.

Materials and equipment

Materials per group

- 2 cm^3 samples of butan-2-ol, 2-methylpropan-2-ol, and 2-methylpropan-1-ol labelled as different unknowns
- iodine solution (10% I_2 in KI(aq)) [dissolve 10 g iodine and 20 g of potassium iodide in deionised water and make up to 100 cm^3]
- 2 mol dm^{-3} sodium hydroxide solution
- 1 mol dm^{-3} dilute sulphuric acid
- 0.1 mol dm^{-3} potassium dichromate(VI) solution [dissolve 2.9 g of potassium dichromate(VI) in deionised water and make up to 100 cm^3].

These reagents are used to perform the iodoform test and to carry out the oxidation of the alcohols.

Equipment per group

- test-tubes
- test-tube rack
- boiling tubes
- 250 cm^3 beaker
- Bunsen burner, tripod, gauze and bench mat
- safety glasses.

Safety

Eye protection must be worn. Students must get their methods checked before they start any practical work. The use of fume cupboards is encouraged.

Risk assessment

A risk assessment must be carried out for this activity.

Commentary

This is an exercise in traditional organic chemistry. A structured approach would be to ask the students to work through the following questions and activities:

(a) Write out the structural formulae of the three possible alcohols.

(b) Classify the alcohols as primary, secondary or tertiary.

(c) Examine the structures and predict which compounds will undergo reactions such as the iodoform reaction or oxidation reactions. There are, of course, other possible reactions. In the case of the two mentioned above you could then ask the students which isomer undergoes: (i) both the oxidation and the iodoform reaction; (ii) the oxidation, but not the iodoform reaction; and (iii) neither reaction.

(d) Carry out the reactions, entering your observations in the following table.

Observations

Alcohol	Iodoform test	Oxidation test
A		
B		
C		

(e) From your observations in (d) give the identity of A, B and C.

Procedures

The following procedures were found to work particularly well in trialling.

Iodomethane reaction

Place six drops of an alcohol in a test-tube. Add 1 cm^3 of iodine in potassium iodide solution followed by sodium hydroxide solution drop by drop until the brown colour of the iodine just disappears (about 2 cm^3). The test is positive if a yellow precipitate (triiodomethane) is produced.

Oxidation reaction

Place dilute sulphuric acid in a boiling tube to a depth of 1 cm. Add three drops of potassium dichromate(VI) solution and then five drops of alcohol. Warm the mixture gently and note if the orange colour of the dichromate(VI) is replaced by the green colour. If this occurs then the alcohol has been oxidised.

Extension

This problem is capable of extension in a variety of ways. During trialling some institutions used different functional group isomers – *eg* $C_3H_6O_2$ ethyl methanoate, methyl ethanoate, propanoic acid; or the three isomers of C_4H_9Cl.

Acknowledgement

This activity is based on a problem written by Joe Burns.

24. Removing the vanadium

Time

3–5 h. This can be used as the basis of a project.

Level

Able A-level students, Higher Grade or equivalent.

Curriculum links

Transition metal chemistry, complex ions. Properties of ionic and covalent substances.

Group size

2.

Materials and equipment

Materials per group

- 0.5 g of $VO(acac)_2$, vanadyl acetylacetonate (available form Aldrich chemicals)
- deionised water
- non-polar organic solvent (*eg* 1,1,1 trichloroethane)
- aqueous solution of sodium sulphide.

Equipment per group

- test-tubes
- test-tube racks
- 100 cm^3 separating funnel
- funnel
- beaker
- safety glasses.

Safety

Eye protection must be worn if practical work is undertaken.

Risk assessment

A risk assessment must be carried out for this activity if practical work is undertaken.

Commentary

This is a real life problem that has been solved successfully. During trialling the problem was both set as a theoretical and practical exercise. The key is to realise that $VO(acac)_2$ is soluble in non-polar solvents and to remove it you must get it into water. To do this the students have to extract the relevant data from the information given. This is not easy but with help most students can achieve it.

One solution to the problem is to wash the reaction mixture with dilute aqueous sodium sulphide. This solution forms a water soluble complex with vanadium and

this is readily demonstrated because the green colour moves between the two immiscible layers.

Vanadyl acetylacetonoate can be made easily.[1]

Extension

Able and interested students could extend this problem by investigating the epoxidation further, and speculate on the role of the vanadium complex in the reaction (the OH group acts as a handle for a complex formed between the vanadium compound and the hydroperoxide: the closest double bond is epoxidised in preference to others). More advanced students may like to speculate as to why NMR spectroscopy doesn't work? (Vanadium is paramagnetic so it broadens the NMR signal.)

Reference

1. M. Fieser and L. F. Fieser, *Reagents for organic synthesis,* **2**, 456. New York, Wiley, 1986. *Idem,* **3**, 331; *Idem,* **11**, 47.

THE ROYAL SOCIETY OF CHEMISTRY

25. The Flatlandian Periodic Table

Time

1–1.5 h.

Level

Able A-level, Higher Grade or equivalent.

Curriculum links

Principle quantum numbers.

Group size

1–3.

Commentary

This problem is challenging! It probes the students' understanding of quantum numbers, bonding and the Periodic Table.

The answers are:

a) The Flatlandian periodic table:

1 $1s^1$											2 $1s^2$
3 $()2s^1$	4 $()2s^2$							5 $()2s^2 2p^1$	6 $()2s^2 2p^2$	7 $()2s^2 2p^3$	8 $()2s^2 2p^4$
9 $()3s^1$	10 $()3s^2$							11 $()3s^2 3p^1$	12 $()3s^2 3p^3$	13 $()3s^2 3p^3$	14 $()3s^2 3p^4$
15 $()4s^1$	16 $()4s^2$	17 $()4s^2 3d^1$	18 $()4s^2 3d^2$	19 $()4s^2 3d^3$	20 $()4s^2 3d^4$	21 $()4s^2 3d^4 4p^1$	22 $()4s^2 3d^4 4p^2$	23 $()4s^2 3d^4 4p^3$	24 $()4s^2 3d^4 4p^4$		

b)

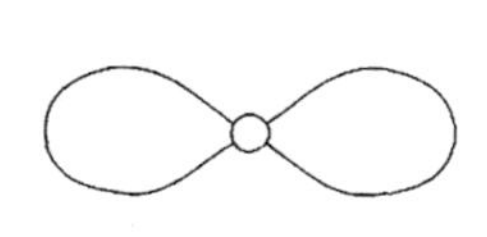

sp^1

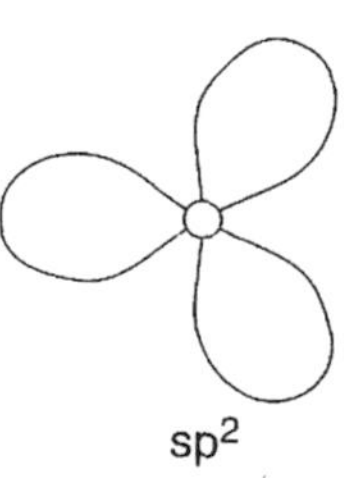

sp^2

The element of life: 5

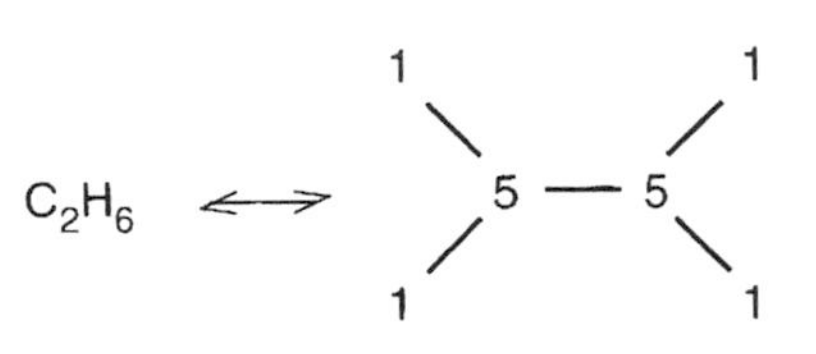

C_2H_4 ⟷ 1 — 5 — 5 — 1

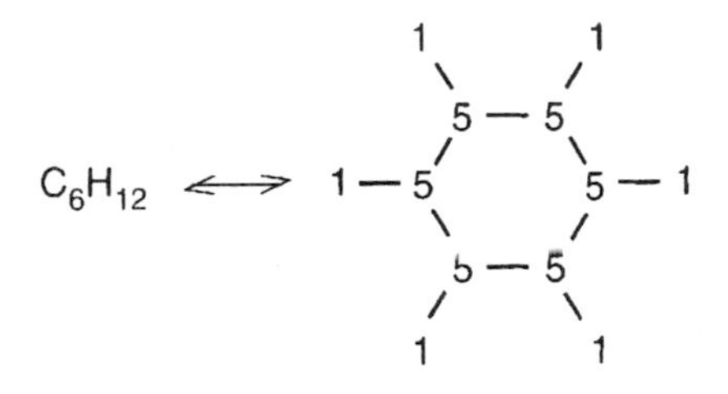

There are no aromatic ring compounds.

c) Sextet rule
14 electron rule

d) The ionisation energies and the trends in electronegativity:

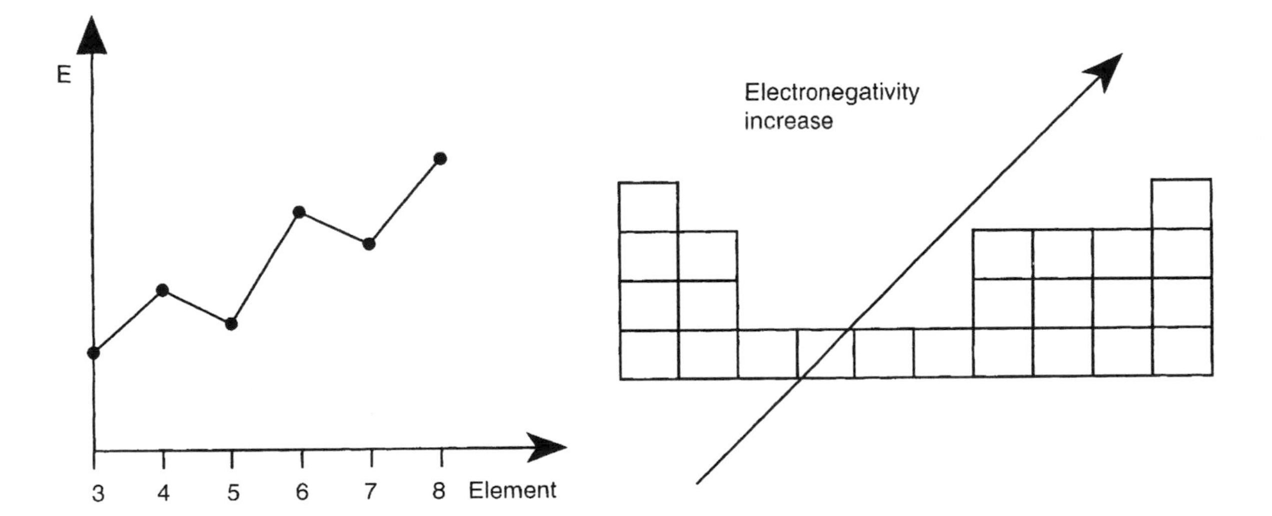

e) The molecular orbital diagram of the homonuclear X_2 molecules:

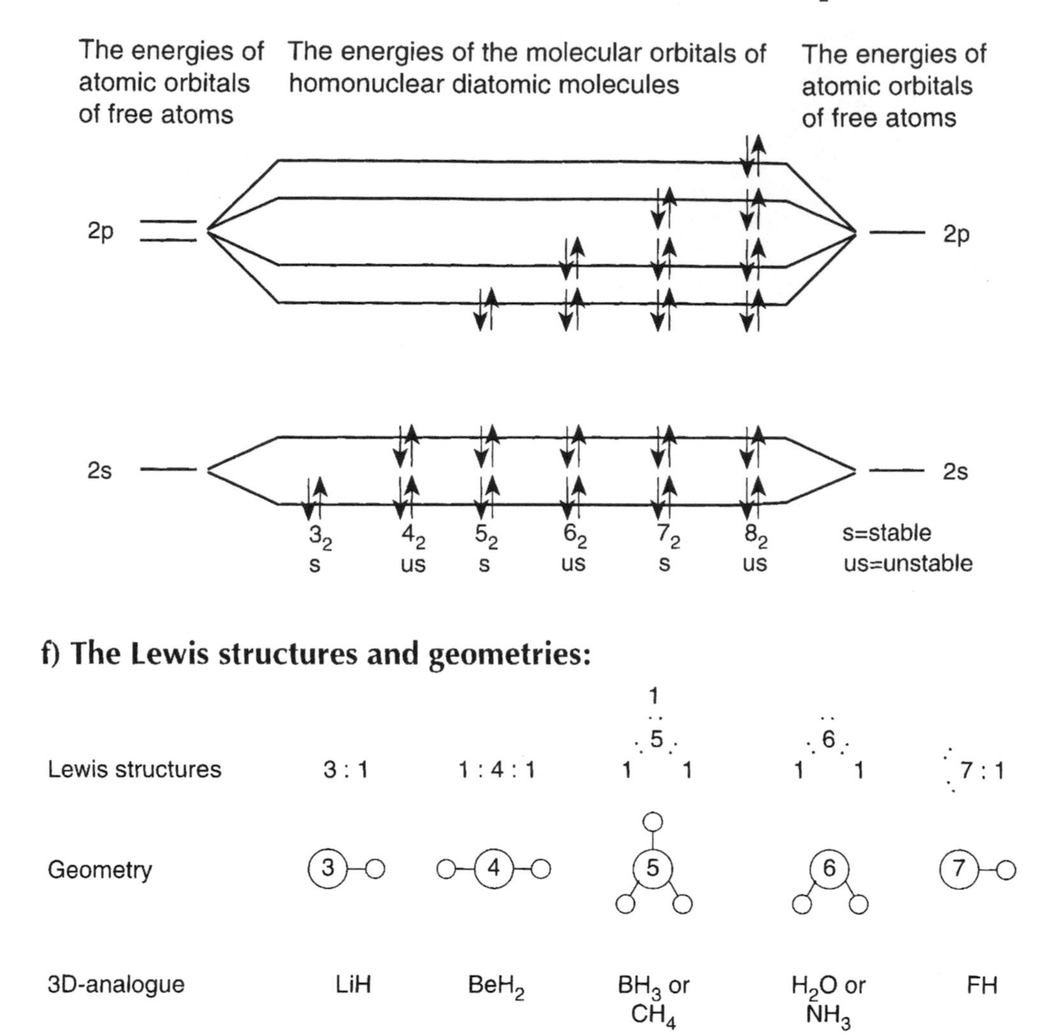

f) The Lewis structures and geometries:

g) The three dimensional analogues and the states of Flatlandian elements:

H					He
Li	Be	B/C	N/O	F	Ne
Na	Mg	Al/Si	P/S	Cl	Ar

G					G
S	S	S	G	G	G
S	S	S	S	G	G

Acknowledgement

The activity is based on a question set at the final of the International Chemistry Olympiad held in Finland during July 1988.

26. Which sodium salt is which?

Time

1–1.5 h.

Level

A-level, Higher Grade or equivalent.

Curriculum links

Tests for sulphur dioxide. Sulphur chemistry – students can find much of this in text books.

Group size

2– 4.

Materials and equipment

Materials per group

1 g of each of the following:

- sodium metabisulphite ($NaHSO_3$)
- sodium sulphate ($Na_2SO_4.10H_2O$)
- sodium peroxodisulphate ($Na_2S_2O_8$)
- sodium sulphite ($Na_2SO_3.7H_2O$)
- sodium thiosulphate ($Na_2S_2O_3.5H_2O$).

- 2 mol dm^{-3} hydrochloric acid
- acidified potassium dichromate(VI) solution
- 0.5 mol dm^{-3} potassium iodide solution
- 0.5 mol dm^{-3} iron(III) chloride solution
- 0.2 mol dm^{-3} iodine in aqueous potassium iodide
- 0.1 mol dm^{-3} silver nitrate solution
- 2 mol dm^{-3} sodium hydroxide solution
- 3% hydrogen peroxide solution
- deionised water
- litmus paper
- strips of filter paper
- safety glasses
- protective gloves.

Equipment per group

- 10 test-tubes
- test-tube rack

Esso

- test-tube holders
- Bunsen burner
- wood splints
- spatula
- wash bottle of deionised water
- dropping pipettes
- safety glasses.

Safety

Eye protection must be worn.

Risk assessment

A risk assessment must be carried out for this activity.

Commentary

Students should be encouraged to predict the reactions of the salts and then draw up a systematic plan for the experiment. The following scheme is set out in the *Independent learning project for advanced chemistry.*[1] It identifies all of the sulphur oxo-anions except the metabisulphite.

Test	Na_2SO_3	Na_2SO_4	$Na_2S_2O_3$	$Na_2S_2O_8$
1. Warm with dilute hydrochloric acid	No reaction in cold. Bubbles of gas on warming. Choking smell. $K_2Cr_2O_7$ paper turned green (SO_2 produced).	No visible reaction.	Solution turned slightly cloudy. Denser when warm. Choking smell. $K_2Cr_2O_7$ turned green (SO_2 produced).	No reactions in cold. Bubbles of gas on warming. Choking smell. Litmus paper bleached (Cl_2 produced).
2. Add silver nitrate solution	Initial white ppt. dissolved on shaking. With more $AgNO_3$ a dense white ppt remained.	No reaction at first, then a faint white ppt. appeared.	Initial white ppt. dissolved on shaking. With more $AgNO_3$ the ppt. remained and turned yellow, brown and black.	Blackish ppt. formed slowly.
3. Add iodine solution (in aqueous potassium iodide)	The brown colour was immediately discharged. (Iodine reduced.)	No visible reaction.	The brown colour was immediately discharged. (Iodine reduced.)	The brown colour become darker. (Iodide oxidised.)
4. Add potassium iodide solution	No visible reaction.	No visible reaction.	No visible reaction.	A dark brown solution was formed. (Iodide oxidised.)

(Continued)

Test	Na_2SO_3	Na_2SO_4	$Na_2S_2O_3$	$Na_2S_2O_8$
5. Add iron (III) chloride solution and dilute acid. Warm and add sodium hydroxide solution.	A dark red-brown solution was formed which became almost colourless when hot. Addition of alkali gave a green ppt. (Fe^{3+} reduced.)	A yellow solution was formed which darkened a little on warming. Addition of alkali gave a red-brown ppt. (Fe^{3+} not reduced.)	A dark purple solution was formed which cleared when hot and then became cloudy. Addition of alkali gave a green ppt. (Fe^{3+} reduced.)	A yellow solution was formed which darkened a little on warming. Addition of alkali gave a red-brown ppt. (Fe^{3+} not reduced.)
6. Heat a small portion of the solid salt.	Crystals turned white and gave off a steamy vapour which condensed on the upper tube (H_2O). White residue turned yellow on strong heating.	A colourless liquid was rapidly formed, which boiled to give off a steamy vapour (H_2O) and a white residue. No further reaction.	A colourless liquid was rapidly formed, which boiled to give off a steamy vapour (H_2O). The yellowish residue turned brown and gave a black viscous liquid.	The solid melted to a colourless liquid. Bubbles of gas relit a glowing splint (O_2).

A procedure for distinguishing between sulphites and metabisulphites is given in Vogel[2] as follows:

Aqueous sulphite shows an alkaline reaction with litmus paper, because of hydrolysis:

$$SO_3^{2-}(aq) + H_2O(l) \longrightarrow HSO_3^-(aq) + OH^-(aq)$$

while aqueous metabisulphite is neutral. On adding a neutral solution of dilute hydrogen peroxide to aqueous sulphite, sulphate ions are formed and the solution becomes neutral:

$$SO_3^{2-}(aq) + H_2O_2(aq) \longrightarrow SO_4^{2-}(aq) + H_2O(l)$$

with aqueous metabisulphite hydrogen peroxide yields hydrogen ions with the same test:

$$HSO_3^-(aq) + H_2O_2(aq) \longrightarrow SO_4^{2-}(aq) + H^+(aq) + H_2O(l)$$

and the solution shows a definite acid reaction.

References

1. *Independent learning project for advanced chemistry, (ILPAC) I 6, Selected p-block elements.* London: John Murray, 1984.

2. Vogel's *Qualitative inorganic analysis,* 6th edn. London: Longman, 1988.

THE ROYAL SOCIETY OF CHEMISTRY

27. What is the smallest amount that you can smell?

Time

1–2 h.

Level

A-level, Higher Grade or equivalent.

Curriculum links

Concentration, solubility of compounds, vapour pressure.

Group size

2– 4.

Materials and equipment

Materials per group

- nice smelling substances – *eg* Muguet, Honeysuckle, Jasmin perfumes (all mixtures) [available from Philip Harris] - or ethoxyethane
- ethanol.

Equipment per group

- 10 and 25 cm^3 measuring cylinders
- dropping pipettes
- 10 cm^3 volumetric flasks
- 1 cm^3 pipettes
- watch glasses
- access to a balance
- safety glasses.

Safety

Eye protection must be worn.

Risk assessment

A risk assessment must be carried out for this activity.

Commentary

There are several different ways of tackling this problem. Each will give a different volume that is the smallest that can be smelt. This volume is also dependent on the individual performing the experiment. In addition, care is needed because continuously smelling the same substance diminishes the sense of smell. If micro-pipettes are available, they can be used directly to give the smallest volume that can be smelt.

A known amount of the perfume can be dissolved in ethanol, and then diluted

until it is just not possible to smell it. Fresh ethanol must be used to ensure an odourless solvent. This is good practice in performing close sequential dilutions. A known volume of the perfume can be placed on a pre-weighed watch glass in the centre of an enclosed room of known volume. When the perfume can be smelt throughout the room (by students already in position in order to minimise air turbulence) then the change in mass of the perfume and the volume of the room can be used to calculate the concentration of perfume vapour.

Extension

Pure fragrances may be used. An example is *cis*-3-hexen-1-ol which smells of cut grass. It is insoluble in water but can be dispersed by using an odourless detergent such as 'Tween 40'.

From data on the vapour pressure of water and the vapour pressure of pure fragrances it is possible to estimate the concentration of fragrance molecules in the air above the aqueous solution.

During trialling it was found that different people have different sensitivities to odours. The differences between people can be a factor of 10 or greater.

28. Which gas is which?

Time

1 h.

Level

A-level, Higher Grade or equivalent.

Curriculum links

Reactions and properties of carbon dioxide, hydrogen, oxygen, chlorine and dinitrogen oxide.

Group size

2– 4.

Materials and equipment

- lime water
- litmus paper
- spills.

Materials per group

Five test-tubes of each of the gases:

- carbon dioxide
- hydrogen
- oxygen
- chlorine
- dinitrogen oxide.

The test-tubes must be inverted and the tops must be under the level of the water.

Equipment per group

- beaker filled with water
- Bunsen burner
- heat resistant mats
- safety glasses.

Safety

Eye protection must be worn if practical work is undertaken.

Risk assessment

A risk assessment must be carried out for this activity if practical work is undertaken.

Commentary

During trialling this was set as both a practical and a theoretical problem. Students looked up the reactions and properties of gases that they were unfamiliar with in order to solve the problem.

Swirling the test-tubes should cause one water level to rise (N_2O). Close examination should yield the yellow colour of chlorine, or a piece of litmus paper held near the bottom of the test-tube should turn red then turn colourless due to bleaching. Removal of the test-tube containing the carbon dioxide and its immersion in a beaker of lime water should give a cloudy solution. By keeping the remaining test-tubes inverted after removing them from the water there should be enough time to carry out the tests for hydrogen and oxygen.

Extension

The experiment can be modified by varying the gases in the test-tubes – *eg* by using nitrogen monoxide and nitrogen dioxide.

29. Identify the metal

Time

1–2 h.

Level

A-level, Higher Grade or equivalent.

Curriculum links

Complex ions. Transition metals.

Group size

2–3.

Materials and equipment

Materials per group

- $Ni(NH_3)_4^{2+}(aq)$ (sometimes regarded as $Ni(NH_3)_6^{2+}(aq)$): dissolve 2 g of $NiSO_4$ in 100 cm^3 water, then add dilute ammonia until the solution turns a deep blue colour.
- $Cu(NH_3)_4^{2+}(aq)$: add ammonia solution dropwise to dilute copper sulphate solution until a deep blue colour is formed (neither concentration is critical).
- $VO^{2+}(aq)$: dissolve ammonium vanadate (V) (ammonium metavanadate, NH_4VO_3) in 2 mol dm^{-3} sulphuric acid to make a 0.1 mol dm^{-3} yellow solution. Add solid sodium metabisulphite until the solution is blue.
- $CoCl_4^{2-}(aq)$: dissolve 0.5 g $CoCl_2.6H_2O$ in conc HCl to give a blue solution. Adding water gradually turns the solution pink, and a purple half-way point can be reached.

For the tests, the following reagents may be useful:

- aqueous sodium carbonate
- aqueous sodium hydroxide
- aqueous ammonia
- conc HCl
- solid buta-2,3-dionedioxime
- aqueous ethylenediamine
- 2-hydroxybenzoate
- 1,2-hydroxybenzoate
- EDTA
- 20 volume hydrogen peroxide
- powdered zinc.

Equipment per group

- test-tubes
- test-tube racks
- dropping pipettes
- full range indicator paper
- safety glasses.

Safety

Eye protection must be worn.
VO^{2+}(aq) must be used in a fume cupboard because a variety of sulphur compounds are produced from the metabisulphite. Alternatively, the blue solution can be boiled for a few minutes to destroy excess metabisulphite.

Risk assessment

A risk assessment must be carried out for this activity.

Commentary

Possible reactions to identify the complexes are:

$Ni(NH_3)_4^{2+}$(aq)

Add dilute hydrochloric acid to the complex solution until it turns green – $Ni(H_2O)_4^{2+}$(aq). To this add:

(a) aqueous sodium carbonate to give green $NiCO_3$(s);

(b) sodium hydroxide solution to give green $Ni(OH)_2$, then add ammonia solution to give blue $Ni(NH_3)_6^{2+}$(aq);

(c) conc HCl to give yellow-green $NiCl_4^{2-}$(aq);

(d) solid buta-2,3-dionedioxime to give a red precipitate; or

(e) the ligand en dropwise to give first a violet solution, $(Ni(H_2O)_6en)^{2+}$, and then a pinky-purple solution, $(Ni(en)_3)^{2+}$(aq).

Other reactions are also possible.

$Cu(NH_3)_4^{2+}$(aq)

Add dropwise:

(a) water to give a pale blue colour, $Cu(H_2O)_4^{2+}$(aq);

(b) EDTA to give a pale blue solution;

(c) conc HCl to give green $CuCl_4^{2-}$(aq);

(d) 2-hydroxybenzoate to give a green solution; or

(e) 1,2-dihydroxybenzoate to give a green solution.

VO^{2+}(aq)

Add to the blue complex solution:

(a) 20 vol hydrogen peroxide to give yellow V^{V}; or

(b) powdered zinc to reduce to first green V^{III} and then lilac V^{II}.

$CoCl_4^{2-}(aq)$

To achieve a blue colour in the preparation of this complex, conc HCl is used. This affects the reactions that might be attempted, and is a clue to the identity of the ligand.

30. Which white salt?

Time

1 h.

Level

A-level, Higher Grade or equivalent.

Curriculum links

Reactions of carbonates. Analysis of ethan-1,2-dioate.

Group size

2– 4.

Materials and equipment

Materials per group

Labelled anhydrous samples of:

- sodium carbonate
- sodium hydrogencarbonate
- sodium ethan-1,2-dioate (oxalate)

and a mixture of the three salts (approximately equivalent molar ratios).

- dilute sulphuric acid
- potassium manganate(VII) solution
- magnesium sulphate
- lime water

and other standard laboratory reagents.

Equipment per group

- normal laboratory glassware
- Bunsen burner and heat resistant mat
- safety glasses.

Safety

Eye protection must be worn.

Risk assessment

A risk assessment must be carried out for this activity.

Esso

Commentary

The most useful tests for identifying the solids are:

1. Oxidation

Sodium ethan-1,2-dioate decolorises a warm acidified potassium manganate(VII) solution; the other two salts do not.

2. Heating

Sodium hydrogencarbonate evolves carbon dioxide with little heat; the residue also evolves carbon dioxide upon treatment with acid

$$2NaHCO_3(s) \longrightarrow Na_2CO_3(s) + H_2O(l) + CO_2(g)$$

Sodium carbonate does not evolve carbon dioxide on heating.

Sodium ethan-1,2-dioate goes black and bubbles on heating, releasing carbon dioxide.

3. Reaction with aqueous magnesium sulphate

Sodium carbonate gives a white precipitate with aqueous magnesium sulphate

$$Na_2CO_3(s) + MgSO_4(aq) \longrightarrow MgCO_3(s) + Na_2SO_4(aq)$$

Sodium hydrogencarbonate gives a white precipitate with aqueous magnesium sulphate only when boiled – initially, soluble magnesium hydrogencarbonate is formed

$$2NaHCO_3(s) + MgSO_4(aq) \longrightarrow Mg(HCO_3)_2(aq) + Na_2SO_4(aq)$$

$$Mg(HCO_3)_2(aq) \longrightarrow MgCO_3(s) + CO_2(g) + H_2O(l)$$

Sodium ethan-1,2-dioate does not give a precipitate with aqueous magnesium sulphate.

Esso

31. Design a pocket handwarmer

Time

1–2 h.

Level

GCSE, Standard Grade or equivalent.

Curriculum links

Simple thermochemical changes.

Group size

2-- 4.

Materials and equipment

Materials per group

- 25 g sodium ethanoate trihydrate
- deionised water

and samples of supersaturated sodium ethanoate solution prepared in advance (see *Demonstration*).

Materials for class demonstration

- 250 g sodium ethanoate trihydrate
- deionised water.

Equipment per group

- boiling water bath or beaker of hot water on a hot plate
- boiling tubes with stoppers
- wash bottles
- heavy duty plastic bags
- small size miscellaneous household items that might be useful in designing the handwarmer – *eg* small plastic bottles or other containers
- 'klippits'
- rubber bands
- paper clips *etc*
- safety glasses.

Equipment for class demonstration

- boiling water bath or beaker of hot water on a hot plate
- large boiling tube or conical flask (with stopper).

Safety

Eye protection must be worn.

Risk assessment

A risk assessment must be carried out for this activity.

Commentary

Handwarmers based on the transfer of heat in various different chemical reactions are available commercially. The best design will be reusable – *ie* it will be possible either to regenerate the reaction or to 'recharge' the device by heating it up.

Possible approach

The 'rechargeable' type requires a closed cycle in which an exothermic process releases heat at a low temperature and the reverse endothermic process takes in heat at a higher temperature. It is suggested that the students are shown the exothermic crystallisation of supersaturated sodium ethanoate described below. The handwarmers sold in many shops use this phenomenon. If the students are encouraged to pursue this approach several samples of supersaturated solutions need to be prepared in advance. The method of initiating recrystallisation will be challenging to students. They should remember that a bump or shock could be sufficient!

Demonstration

Crystallisation from supersaturated solutions of sodium ethanoate

1. To 250 g of sodium ethanoate trihydrate in a large boiling tube add 100 cm^3 of deionised water.
2. Set up a boiling water bath or set up a large beaker of boiling water on a hot plate.
3. Heat the mixture in the water bath with occasional swirling until a clear solution is obtained.
4. Using a wash bottle, carefully rinse the glass surface at the top of the tube.
5. Insert a stopper and allow the tube and its contents to cool to room temperature (this will take 1–3 hours). The process can be speeded up by placing the tube in a large beaker and cooling it with running water.
6. To start recrystallisation remove the stopper and carefully drop a single crystal of sodium ethanoate trihydrate into the tube. Crystallisation occurs with the evolution of heat.
7. The tube can be reheated and the process repeated.

An alternative approach[1] involves a mixture of iron powder, sodium chloride, vermiculite and water in a closed plastic bag. An exothermic reaction occurs as the iron is oxidised. The reaction can be controlled by restricting the amount of air entering the bag.

Extension

1. The design of cold packs could also be considered. These can be based on the absorption of heat when ammonium nitrate dissolves in water.
2. The reaction described in the *Commentary* could be the basis of a chemical heat pump.

Reference

1. L. R. Summerlin, C. L. Borgford, and J. B. Ealy, *Chemical demonstrations: A sourcebook for teachers, Vol 2.* Washington: ACS, 1987.

Acknowledgement

This activity is based on a suggestion by Karen Davies.

32. A hot dinner from a can

Time

1–2 h.

Level

GCSE, Standard Grade or equivalent.

Curriculum links

Simple thermochemistry and enthalpy calculations.

Group size

2– 4.

Materials and equipment

Materials per group

- fresh calcium oxide (approximately 50–100 g per group). The calcium oxide must not be slaked and fresh commercial CaO is recommended; alternatively CaO can be obtained by roasting $CaCO_3$ in a kiln (some institutions may have a Muffle furnace). The temperature required is over 1000°C so a Bunsen burner is not hot enough. The CaO should be tested before the session to avoid disappointment.

Equipment per group

Items from the junk list (pXX), including types of insulation

- glass stirring rods
- glass beakers
- boiling tubes
- test-tubes
- thermometers
- safety glasses.

Safety

Eye protection must be worn.

The reaction can be unpredictable (probably because $Ca(OH)_2$ forms on the surface of the CaO). In one case a student added 5 cm^3 of water to about 50 g of CaO all at once and the evolution of heat was very rapid – cracking the pyrex beaker quite violently. Care should be taken in disposing of unreacted CaO as a delayed exothermic reaction could occur in the waste pipe if it is accidentally washed down the sink!

Risk assessment

A risk assessment must be carried out for this activity.

Commentary

It is suggested that the students use the exothermic reaction between calcium oxide and water to produce a steady supply of heat.

$$CaO(s) + H_2O(l) \longrightarrow Ca(OH)_2(s)$$

The students will need to calculate the quantities of heat involved before designing the system. The enthalpy change for the reaction above is approximately –65 kJ mol^{-1}. This information may be provided or they may calculate it by using standard molar enthalpy changes obtained from a data book. They should then find the heat transferred when 1 g of calcium oxide reacts with water. During trials the students were encouraged to do small-scale trial runs and this worked particularly well. The specific heat capacity of the food may be approximated to that of water for the purposes of this activity.

An important criterion in assessing the final design is that the system should be as compact and light as possible. 'Hotcans' that work on this principle can be bought in camping shops and the best student designs could be compared with the commercial product.

Extension

The students could be asked to consider how the proposed device would perform under adverse weather conditions – *ie* would it work below freezing point?

33. A chemical stop-clock: iodine clock reaction

Time

1 h.

Level

GCSE, Standard Grade or equivalent.

Curriculum links

Rates of reaction.

Group size

1–2.

Materials and equipment

Materials per group

These solutions are better made up fresh, not more than 24 h before they are required.

- Solution A: 2.1 g potassium iodate(V) is dissolved in 1 dm^3 deionised water followed by the addition of 10 cm^3 of 1 mol dm^{-3} sulphuric acid.
- Solution B: 4 g of soluble starch is made up into a paste with a little cold water and this is added to 1 dm^3 boiling deionised water. 0.9 g of sodium hydrogensulphite and 10 cm^3 of 1 mol dm^{-3} sulphuric acid are added to the cooled solution. (Sodium hydrogensulphate is available as a solution from some suppliers.)

Equipment per group

- two 100 cm^3 beakers
- two 250 cm^3 beakers
- 25 cm^3 measuring cylinder
- two 100 cm^3 measuring cylinders
- two 250 cm^3 measuring cylinders
- white tile
- stirring rod
- stop-watch
- the use of burettes allows more accurate measurements to be made
- safety glasses
- graph paper.

If this problem is used in a competition then a large display digital clock can heighten the excitement at the final stage.

THE ROYAL
SOCIETY OF
CHEMISTRY

Safety

Eye protection must be worn.

Risk assessment

A risk assessment must be carried out for this activity.

Commentary

This problem can be approached as a competition, in which case the time available for experimenting should be limited. Sufficient time must be left at the end for judging.

Extension

The competition may be made more difficult by limiting the volume of stock solutions available to each competitor.

Acknowledgement

This activity is based on a problem used at Norwich Chemical Olympiad in 1984.

THE ROYAL
SOCIETY OF
CHEMISTRY

34. Lifting an egg by a thread

Time

1–1.5 h.

Level

GCSE, Standard Grade or equivalent.

Curriculum links

The more problem solving the students have done the better.

Group size

2–3.

Materials and equipment

Items from the junk list (pXX) – this must include items that allow the gel to be extruded *ie* syringes, cake icer, washing-up bottles.

Materials per group

- 300 cm^3 of 880 ammonia solution
- 30 g powdered copper carbonate
- 6 g cellulose (filter paper, cotton wool)
- 1 mol dm^{-3} sulphuric acid
- small fresh egg
- deionised water.

Equipment per group

- two 250 cm^3 beakers
- glass rod
- shallow plastic tray (to extrude the cellulose into)
- safety glasses.

Safety

Eye protection must be worn. Note that 880 ammonia is an irritant.

Risk assessment

A risk assessment must be carried out for this activity

Commentary

This problem caused much consternation during trialling because it is more difficult than it seems. The trick is to get the recipe right and to make fairly thick fibres. These will lift an egg if used carefully although the fibres degrade after a while. Using cotton wool as the source of cellulose was found to give the best results. In the trials one group cheated first by using a blown egg, and then a small pigeon egg: both were spotted and banned!

Procedure

The following procedure was found to give good results.

10 g of copper carbonate was added to 100 cm^3 of 880 ammonia solution in a beaker, until no more dissolved. After two minutes the blue solution was decanted off. 1 g of finely shredded cotton wool was stirred in gently, taking extreme care not to fold in air, until the blue solution has the consistency of a shower gel. Between 1 and 1.5 g of cellulose was needed. Cellulose fibres were reformed by extruding the solution using a 20 cm^3 syringe into a 1 cm depth of 1 mol dm^{-3} sulphuric acid solution in a tray. The fibres can be washed with water after 20–40 minutes. The ideal fibres are *ca* 2–3 mm thick and the egg can be lifted gently by using the fibres to form a cradle.

Acknowledgement

This activity was based on an idea by John Crellin. The procedure outlined was developed by Valerie Tordoff at Eton College.

35. Number of reactions

Time

1 h.

Level

GCSE, Standard Grade or equivalent.

Curriculum links.

Chemistry of copper. Weak bases.

Group size

2– 4.

Materials and equipment

Materials per group

- 0.25 mol dm^{-3} aqueous copper(II) sulphate
- 0.25 mol dm^{-3} aqueous ammonium carbonate labelled A.

Equipment per group

- test-tubes
- test-tube racks
- glass rods
- pH paper
- lime water
- safety glasses.

Safety

Eye protection must be worn.

Risk assessment

A risk assessment must be carried out for this activity.

Commentary

This is an open problem and requires the students to exercise care and make careful observations. One approach is to look at solution A first. On warming, this solution evolves ammonia which gives the characteristic smell and carbon dioxide.

On mixing solution A with aqueous copper sulphate a blue/green precipitate ($Cu(OH)_2.CuCO_3$) is formed initially. On adding excess A this dissolves to form a deep blue solution of $Cu(NH_3)_4^{2+}(aq)$.

Evaluation of solution

The best answers are those that identify A, and identify the correct reactions. Some of these can be written as:

$(NH_4)_2CO_3(aq) \rightleftharpoons 2NH_3(aq) + \text{"}H_2CO_3\text{"}(aq)$

$(NH_4)_2CO_3(aq) \rightleftharpoons 2NH_4^+(aq) + CO_3^{2-}(aq)$

$NH_4^+(aq) + H_2O(l) \rightleftharpoons NH_3(g) + H_3O^+(aq)$

$CO_3^{2-}(aq) + H_2O(l) \rightleftharpoons HCO_3^-(aq) + OH^-(aq)$

$\text{"}H_2CO_3\text{"}(aq) \rightleftharpoons H_2O(l) + CO_2(g)$

$Cu^{2+}(aq) + 2OH^-(aq) \rightleftharpoons Cu(OH)_2(s)$

$Cu^{2+}(aq) + CO_3^{2-}(aq) \rightleftharpoons CuCO_3(s)$

$Cu^{2+}(aq) + 4NH_3(aq) \rightleftharpoons Cu(NH_3)_4^{2+}(aq)$

$Ca(OH)_2(s) + CO_2(g) \longrightarrow CaCO_3(s) + H_2O(l)$

$CaCO_3(s) + H_2O(l) + CO_2(g) \longrightarrow Ca(HCO_3)_2(aq)$

36. Gas volume

Time

1–2 h.

Level

GCSE, Standard Grade or equivalent.

Curriculum links

Reactions of metals with acids. Moles.

Group size

2– 4.

Materials and equipment

General

- a range of laboratory glassware including measuring cylinders, burettes and beakers
- access to two decimal place balance.
- items from the junk list (pXX).

Materials per group

- bottle of vinegar
- magnesium ribbon cut into 0.04g pieces.

Safety

Eye protection must be worn.

Risk assessment

A risk assessment must be carried out for this activity.

Commentary

There are many approaches to this problem but they all share a common aim to measure the volume of hydrogen evolved from a known mass of magnesium. Most approaches collect the gas over water in a graduated device such as a measuring cylinder or burette. Students need to consider the accuracy of the method that they choose. The volume of gas is noted and adjusted for atmospheric pressure. It is then simple to calculate the volume of 1 mole of the gas.

$$Mg(s) + 2CH_3COOH(aq) \longrightarrow (CH_3COO)_2Mg(aq) + H_2(g)$$

1 mole 1 mole

Evaluation of solution

This problem has been used as a competition. The best solution is the one that is nearest the theoretical value but credit should also be given for elegance. During trialling many students forgot to adjust the volume of gas collected to atmospheric pressure and room temperature – *ie*

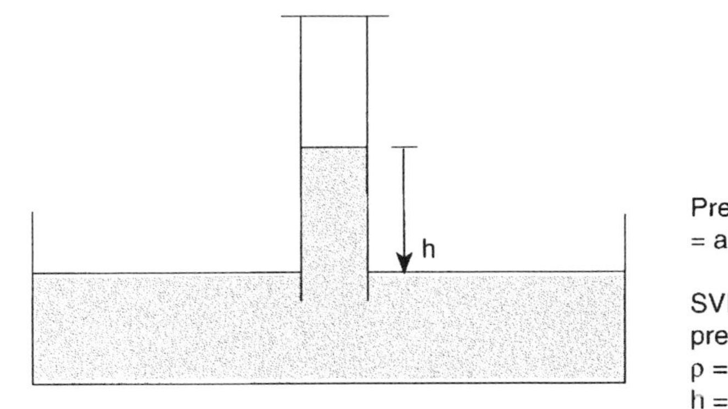

Pressure of gas + ρgh + SVP
= atmospheric pressure

SVP = Saturated vapour pressure of water
ρ = density of water
h = height of water

The corrections are very small. The difference between theoretical and experimental answers is usually very small.

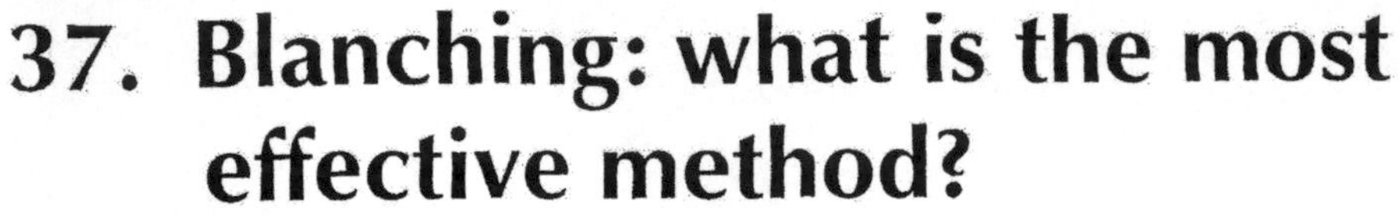

37. Blanching: what is the most effective method?

Time

1–2 h.

Level

GCSE, Standard Grade or equivalent.

Curriculum links

Enzymes, catalysts.

Group size

2– 4.

Materials and equipment

Materials per group

- 1 kg of Brussels sprouts
- 0.5% v/v hydrogen peroxide solution
- 1% guaiacol solution
- access to deionised water.

Equipment per group

- teat pipettes
- large beakers
- Bunsen burner
- sieve
- knife
- access to balance
- items from junk list (pXX) to make grading device
- safety glasses.

Safety

Eye protection must be worn.

Risk assessment

A risk assessment must be carried out for this activity.

Commentary

This problem mirrors the process that is used in the food industry. Students should realise that the first task is to sort the sprouts into grades, three are sufficient. During trialling two square holes were used for the grading: 18 mm and 22 mm. Sprouts

passing through the small hole were graded as small, while those that did not pass through the 22 mm hole were large; the remainder were medium. The sprouts should be blanched in boiling water and it should be kept boiling during the blanching. Trialling showed that approximately 5 minutes is required although the time depends on size, and the same water can be used repeatedly for the blanching. However it is important that a fresh water cooling solution is used each time. Cooling should take around 5 minutes.

Adding 3 drops of each of the hydrogen peroxide and guaiacol solution to cut sprouts should give an immediate result. If a brown colour appears within 10 s then this is a positive test for peroxidase activity. If colour appears only after 10 s or does not appear then this is a negative result. During trialling it was noted that a rapid colour change indicated severe underblanching, a slower colour change may indicate that blanching is about right, while a lack of any colour change after several minutes may indicate overblanching. Students who have difficulty with the test should try it out on unblanched sprouts first.

Extension

This problem has been set as a competition. Students were asked to estimate the relative cost of blanching each fraction size; to plot blanching time against size; and to suggest what other vegetables could be used.

Broccoli and carrots can be used as alternative vegetables, and these do not need grading. They must, however, be cut up into uniform sized pieces. Mange-tout (or sugar-snap peas) can also be used, but need to be ground together with calcium carbonate and test for an alternative enzyme, catalase.

Acknowledgement

This activity is based on an exercise used at a regional heat for 'Top of the Bench' competition. The Society is grateful to Chris Harbord of Birds Eye Walls for his advice.

38. Cool it

Time

1 h.

Level

12 years and upward.

Curriculum links

Chemical reactions that involve temperature changes.

Group size

2– 4.

Materials and equipment

Materials per group

- deionised water
- 10 g citric acid
- 10 g sodium hydrogencarbonate.

Equipment per group

- polystyrene cup
- –10 to +100°C thermometer
- stirring rod
- stop clock
- spatula
- two 50 cm^3 measuring cylinders
- safety glasses.

Safety

Eye protection must be worn.

Risk assessment

A risk assessment must be carried out for this activity.

Commentary

The reaction produces an effect similar to that of sherbet in the mouth. It is an endothermic reaction and the best results are obtained when a slurry is used. Students need to ensure that their results are reproducible. Younger students may need help in choosing the volume of water to be measured.

Evaluation of solution

This problem has been used as a competition. Judges measured the temperature fall 1 minute after the reaction started. The order of merit depended on how close the temperature was to the target and how quickly teams were ready to be judged.

Extension

The temperature given in the problem can be changed; 10.5°C would have served just as well.

39. Liesegang rings: how the tiger got its stripes

Time

The formation of Liesegang rings is a slow process (taking hours to days) but the experiments are simple to set up. This investigation is therefore more suitable to project work undertaken over a few weeks.

Level

12 years and upwards.

Curriculum links

Precipitation reactions.

Group size

2– 4.

Materials and equipment

Materials per group

- 20 g gelatin
- deionised water
- 2.5 g cobalt(II) chloride
- 2.5 g magnesium chloride
- 2.5 g copper(II) sulphate
- 2.5 g manganese(II) chloride
- 2.5 g copper(II) chloride
- 2.5 g potassium chromate(VI)
- 1.0 g silver nitrate
- 20 cm^3 of 880 ammonia
- 20 cm^3 of 19 mol dm^{-3} sodium hydroxide
- 20 cm^3 of 0.1 mol dm^{-3} silver nitrate solution.

Equipment per group

- test-tubes
- test-tube racks
- parafilm
- safety glasses.

Safety

Eye protection must be worn.

Risk assessment

A risk assessment must be carried out for this activity.

Commentary

Liesegang rings were the subject of many studies at the turn of the century.[1] The production of the rings or bands can be a valuable project for chemistry students as the investigations are open-ended[2] and it is possible to design many different experiments using simple apparatus. The phenomena are fascinating and highly variable. Flicker and Ross[3] have discussed some of the theories that have been proposed to explain the rings, but there is no universally accepted theory to account for their formation. The position of the student is much closer to that of a practising scientist than when the 'right answer' is known from a secondary source.

It is now recognised that Liesegang rings are an example of an oscillating chemical reaction.[4] There has been an explosion of interest in these reactions, which began with what is known as the Belousov-Zhabotinsky (BZ) reaction. Chemical oscillations can take different forms – *eg* if a reaction mixture is stirred, periodic variations in time may be observed. For instance, one stirred BZ solution constantly changes from red to blue and back. In the absence of stirring, the concentrations may vary from place to place as well as through time. The oscillation becomes a travelling wave and can give rise to complex patterns. The BZ solution will give these effects if the solution forms a very shallow, unstirred layer.[5] Liesegang rings are rhythmic precipitation patterns occurring within a gel.

In this process of self organisation, order is appearing spontaneously from disorder. Chemists were very reluctant to accept this idea. Their resistance was rooted in popular but erroneous statements of the Second Law of Thermodynamics to the effect that, if a process occurs spontaneously, entropy or disorder tends to increase. Such statements are incomplete because classical thermodynamics also requires that the system should be near its equilibrium state. Oscillating reactions occur in systems that are far from equilibrium. Ilya Prigogine developed the theory of thermodynamics to include far from equilibrium systems and he showed that ordered structures can develop from disorder in these systems and was awarded the Nobel Prize for chemistry in 1977 for his contribution to thermodynamics.[6]

The mathematician Alan Turing showed that oscillating reactions were theoretically possible. He showed that the physical laws governing the reactions and diffusion of chemical substances could explain the way in which pattern and structure might develop in living things.[7] The physical explanation of the beautiful patterns that develop in the gels has features in common with the way a tiger gets its stripes.

Procedure

It is suggested that the students should first follow the preliminary instructions provided below. When they have set up systems which show the effects they should be asked to put forward hypotheses about the phenomena and to design experiments to test them.

Two sets of instructions are provided. The concentrations differ as the experiments come from different sources; the first set is based on a method for producing 'very perfect silver chromate rings' described in *The science masters' book*.[8]

THE ROYAL SOCIETY OF CHEMISTRY

Preliminary instruction

Preparation of a gel containing potassium dichromate(VI) covered by silver nitrate solution

(a) Weigh out 2.5 g of gelatin and 0.025 g of potassium dichromate(VI). Add 50 cm^3 of deionised water. Heat gently, with stirring, until the solution is clear. Pour into test-tubes of various diameters so that the tubes are about two thirds full. Cover the test-tube with parafilm and leave the solution to set.

(b) Weigh out 0.85 g of silver nitrate and dissolve in 10 cm^3 deionised water. About 1 cm^3 of this solution is poured on top of the gel. The test-tube is then covered with parafilm and left undisturbed.

Over the next few days the formation of bands of colour will be observed. The experiment may be modified by pouring a small amount of the gelatin-potassium dichromate(VI) solution onto a glass slide or into a crystallising dish and, after it has set, dropping on a very small amount of silver nitrate solution. Concentric rings will be seen to develop.

Preparation of a gel containing cobalt(II) chloride covered by 880 ammonia

Weigh out 1.5 g of gelatin and 2.5 g of cobalt(II) chloride and add 50 cm^3 of deionised water. Heat gently with stirring until the solution is clear. Pour the solution into a test-tube (25 x 150 mm) until it is about two-thirds full. Cover the tube with parafilm and leave undisturbed until a gel is formed.

Carefully pour 880 ammonia on top of the gel until the tube is almost full. Cover the tube with parafilm. The tube should be left to stand for a few days when bands will be observed to form.

This is the basic procedure suggested by Schibeci and Carlsen.[2] They tried out a variety of systems to produce Liesegang bands and obtained successful results with the following:

Gel containing	Solution on top
cobalt(II) chloride	880 ammonia
magnesium chloride	sodium hydroxide (19 mol dm^{-3})
magnesium chloride	880 ammonia
copper(II) sulphate	silver nitrate (0.1 mol dm^{-3})
manganese(II) chloride	880 ammonia
uranyl nitrate	880 ammonia
uranyl nitrate	sodium hydroxide (19 mol dm^{-3})
uranyl nitrate	silver nitrate (0.1 mol dm^{-3})
copper(II) chloride	sodium hydroxide (19 mol dm^{-3})
potassium chromate(VI)	silver nitrate (0.1 mol dm^{-3})

The article by these authors[2] is particularly helpful because they also give combinations which, they found, did not give the banded effects.

Extension

Further lines of investigation could include the following.

1. Obtaining a three dimensional effect. Liesegang[1] describes patterns like the skins of an onion forming when a block of dichromate gelatin is placed in silver nitrate solution.
2. Varying the concentrations of the reagents.
3. Adding a mixture of substances to the gel – *eg* cobalt(II) chloride, magnesium chloride and manganese(II) chloride.
4. Using test-tubes of differing diameter and glassware of different shapes.
5. Replacing the gelatin with another material to give a gel: Flicker and Ross describe an experiment using agar.[3] Sharbaugh and Sharbaugh describe, in a beautifully illustrated article,[9] their researches using gels prepared by mixing equal volumes of water glass or sodium silicate (density 1.06 g cm^{-3}) and 0.5 mol dm^{-3} ethanoic acid.
6. Looking at geometrical relationships in the patterns formed: is the spacing between the bands uniform? Are the bands rings or spirals?
7. Discovering what happens to the bands when they reach the bottom of the tube.

References

1. R. E. Liesegang, *Chemische reactionen in gallerten*. Dresden und Leipzig: Theodor Steinkopff, 1924.
2. R. A. Schibeci and C. Carlsen, *J. Chem. Ed.*, 1988, **65**, 365.
3. M. Flicker and J. Ross, *J. Chem. Phys.*, 1974, **60**, 3458.
4. I. R. Epstein, K. Kustin, P. De Kepper and M. Orban, *Sci. Am.*, 1983, **283**, 96.
5. P. W. Atkins, *Atoms, electrons, and change*. New York: Scientific American Library, 1991.
6. P. Coveney and R. Highfield, *The arrow of time*. London: W. H. Allen, 1990.
7. A. M. Turing, *Philos. Trans. R. Soc. B*, 1952, **237**, 37.
8. *The science masters' book*, part II, series 1.
9. A. H. Sharbaugh and A. H. Sharbaugh, *J. Chem. Educ.*, 1989, **66**, 589.

40. The candle in the bell-jar

Time

2 h or longer.

Level

12 years and upwards.

Curriculum links

Hydrocarbon chemistry. Combustion.

Group size

2– 4.

Materials and equipment

Materials per group

- candles (the students may decide that they would like to experiment with different heights and thicknesses).

Equipment per group

- bell jar or large glass beaker
- beehive shelf
- glass trough
- matches
- general laboratory ware
- safety glasses.

Safety

Eye protection must be worn.

Risk assessment

A risk assessment must be carried out for this activity

Commentary

The first part of this challenge is to work out the three hypotheses given in the *Student Sheet* for this phenomenon:

(1) the rise in the water level is caused by the consumption of oxygen alone;

(2) the rise in the water level is caused by a combination of the oxygen consumed and the carbon dioxide released; and

(3) the rise in the water level is caused by the contraction of gases in the bell jar as they cool.

Each of these hypotheses gives rise to several lines of investigation. Students could, for instance, design a system for lighting the candle inside an enclosed volume

of air to give the most accurate assessment of the rise in the water level. They might try introducing a layer of oil above the water to prevent carbon dioxide from dissolving, to see if this affects the rise in the water level. The candle could be replaced by a spirit burner so that the amount of fuel burned can be used to calculate the volume of oxygen consumed (and carbon dioxide released). An alternative method of heating the gases inside the bell jar to the same temperature as the lighted candle can be tried, to assess the rise in water level on cooling.

A critical appraisal of the experiment was published in 1967 by Richard Kempa[1] who showed that the combustion of a candle cannot be supported by an atmosphere containing less than about 14 per cent oxygen by volume. He suggested that the experiment may be used to introduce children to the idea that oxygen does not support combustion if it is insufficiently "concentrated".

Reference

1. R. F. Kempa, *Science teaching techniques 12.* London: John Murray, 1967.

Acknowledgement

This activity is based on a suggestion from John Barker.

THE ROYAL SOCIETY OF CHEMISTRY

41. Move an Oxo cube at great speed

Time

It is suggested that either:

an entire morning be devoted to the problem (*eg* on the last day of term), which would allow 2 h for practical activities and 30 minutes for judging

or

the problem be given to the class as a homework exercise 2 weeks or so before the judging. Judging could then take place in a normal double science lesson, allowing 45 minutes for repair and final adjustments, and 30 minutes for judging.

Level

12 years and upward.

The exercise is better as a pre-set problem for younger students.

Curriculum links

Production of carbon dioxide.

Group size

3– 4.

Materials and equipment

- items from the junk list (pXX) to encourage creativity
- the judges will require a stop watch or an arrangement with a scaler and photocells if the Oxo cubes move too fast.

Materials per group

- sodium hydrogencarbonate (maximum amount = 3 level teaspoons)
- citric acid (maximum amount = 9 level teaspoons)
- access to water
- butter/margarine to reduce friction.

Equipment per group

- identical teaspoons (can be plastic)
- safety glasses.

Safety

Eye protection must be worn.

Risk assessment

A risk assessment must be carried out for this activity.

Commentary

Some guidance may be needed for younger age groups – *eg* water is needed for the reaction, or use 'Andrews'. The reaction might be used to do the moving, or it could be used to start the movement – *eg* to trigger movement of a counterbalance.

Evaluation of solution

These are suggestions only.

1. The final device should be loaded with chemicals, and be ready to start when the judge says so.
2. The judge will provide each group with the levelled teaspoons of chemicals for the test. (Judges may prefer to weigh out the relevant amounts.)
3. The winner is the team whose device moves the Oxo cube over the course in the shortest time.
4. In the event of a tie, the judge should take into account the elegance of the solution, given the requirement that the devices are constructed mainly from junk materials.

Extension work

To increase the chemical content the task could be extended by prior (or subsequent) experimentation, to select best choice of gases/chemicals.

Acknowledgement

This activity is based on an idea by Peter Borrows.

42. Move a heavy object

Time

It is suggested that either:

an entire morning be devoted to the problem (*eg* on the last day of term), which would allow 2 h for practical activities and 30 minutes for judging

or

the problem be given to the class as a homework exercise 2 weeks or so before the judging. Judging could then take place in a normal double science lesson, allowing 45 minutes for repair and final adjustments, and 30 minutes for judging.

Level

12 years and upward.

The exercise is better as a pre-set problem for younger students.

Curriculum links

Production of carbon dioxide gas.

Group size

3– 4.

Materials and equipment

The judges will require a stop watch.

Materials per group

- sodium hydrogencarbonate (maximum amount = 3 level teaspoons)
- citric acid (maximum amount = 9 level teaspoons)
- access to water
- butter/margarine to reduce friction.

Equipment per group

- items from the junk list (pXX) to encourage creativity
- identical teaspoons (can be plastic)
- safety glasses.

Safety

Eye protection must be worn.

Risk assessment

A risk assessment must be carried out for this activity.

Commentary

During trialling guidance needed to be given to younger age groups to say that water is needed for the reaction, or use 'Andrews'. The reaction might be used to do the moving, or it could be used to start the movement, *eg* to trigger movement of a counterbalance.

Evaluation of solution

These are suggestions only.

1. The final device should be loaded with chemicals, and be ready to start when the judge says so.
2. The judge will provide each group with the levelled teaspoons of chemicals for the test. (Judges may prefer to weigh out the relevant amounts.)
3. The winner is the team whose device moves the heaviest object over the course.
4. In the event of a tie, the judge should take into account the elegance of the solution, given the requirement that the device shall be constructed mainly from junk materials.

Extension work

To increase the chemical content the task could be extended by prior (or subsequent) experimentation, to select best choice of gases/chemicals.

Acknowledgement

This activity is based on a suggestion by Peter Borrows.

43. Test the gas

Time

1 h.

Level

12 years and upward.

Curriculum links

Test for carbon dioxide.

Group size

2–4.

Materials and equipment

Materials per group

- gas jar full of carbon dioxide – set up as below
- fresh lime water.

Equipment per group

- various lengths of tubing
- plastic syringes
- items from the junk list (pXX)
- plastic bowl
- safety glasses.

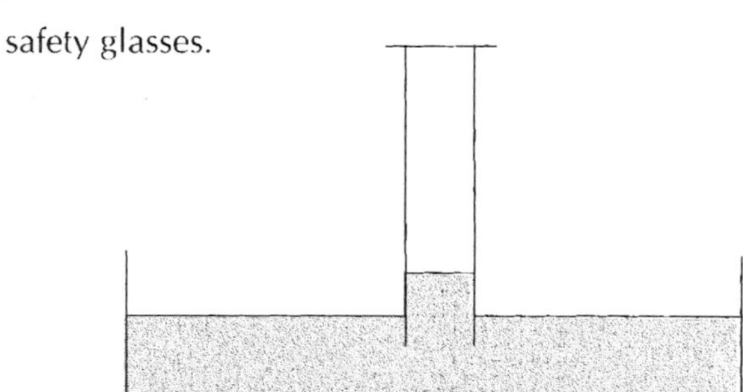

Safety

Eye protection must be worn.

Risk assessment

A risk assessment must be carried out for this activity.

Commentary

There are many approaches to this problem, some more ingenious than others. The essential idea is to construct something that will go under the water and collect the gas. During trialling students designed containers filled with water, which when under the gas jar could be emptied and the gas collected. Syringes connected to rubber tubes were banned! A time element can be introduced to focus the activity.

Extension

The experiment can be modified by varying the gas in the gas jar.

44. What makes the candle go out?

Time

1 h.

Level

12 years and upward.

Curriculum links

Control of variables. Hypothesis testing.

Group size

2– 4.

Materials and equipment

Materials per group

- candles of various sizes and of different types.
- matches.

Equipment per group

- heat resistant mats
- gas jars
- safety glasses.

Safety

Eye protection must be worn.

Risk assessment

A risk assessment must be carried out for this activity.

Commentary

This activity is designed to encourage students to take a logical approach in solving a problem with many variables. After initial investigation, the student should realise that the height of the candle, the type of wick and the type of wax may all influence the results. Therefore a proper conclusion can only be drawn if all the variables, except the one under test, are kept constant. Remember to remind students to check for air leaks and mechanical flaws in the apparatus.

Results show that if two or more similar candles of different heights are used, the tallest one goes out first. If two candles of the same height but different wick thicknesses are used the candle with the thinnest wick goes out first. The type of candle wax is also a factor.

The scientific explanation for these results is unknown!

Extension

This activity has been extended to include three candles of different heights. It was difficult to obtain consistent results, but we don't understand why!

45. Warning device

Time

It is suggested that either:

an entire morning be devoted to the problem (*eg* on the last day of term), which would allow 2 hours for practical activities and 30 minutes for judging

or

the problem be given to the class as a homework exercise 2 weeks or so before the judging. Judging could then take place in a normal double science lesson, allowing 45 minutes for repair and final adjustments, and 30 minutes for judging.

Level

12 years and upward.

The exercise is better as a pre-set problem for younger students.

Curriculum links

Production of carbon dioxide gas.

Group size

3– 4.

Materials and equipment

Items from the junk list, for example balloons, surgical gloves, (pXX) – should be chosen to encourage creativity.

Materials per group

- sodium hydrogencarbonate (maximum amount = 3 level teaspoons)
- citric acid (maximum amount = 9 level teaspoons)
- access to water.

Equipment per group

- identical teaspoons (can be plastic)
- safety glasses.

Safety

Eye protection must be worn.

Risk assessment

A risk assessment must be carried out for this activity.

Commentary

Guidance may be needed for younger age groups to say that water is needed for the reaction, or use 'Andrews'. The reaction might be used to do the moving, or it could be used to start the movement – *eg* to trigger the movement of a counterbalance to light a bulb or rattle a tin. During trialling some students inflated a surgical glove while others used the reaction to complete an electric circuit and thus rung a bell. A variety of approaches were seen – some more elegant than others!

Evaluation of solution

These are suggestions only.

1. The final device should be loaded with chemicals, and be ready to start when the judge says so.
2. The judge will provide each group with the levelled teaspoons of chemicals for the test. (Judges may prefer to weigh out the relevant amounts.)
3. The winner is the team whose device gives the best warning.
4. In the event of a tie, the judge should take into account the elegance of the solution, given the requirement that the device shall be constructed mainly from junk materials.

Extension

To increase the chemical content the task could be extended by prior (or subsequent) experimentation, to select the best choice of gases/chemicals.

46. An elementary problem

Time

1 h.

Level

12 years and upward.

Curriculum links

Industrial processes.

Group size

1–3.

Commentary

This problem is easier if you ignore the chemistry and treat it as a logic problem. Students enjoyed doing it, as did the teachers! If students are struggling to get started encourage them to draw up a grid with the variables across the top and down the side. This will enable them to mark in correlations.

Answer

The answer is

Company	Chemical	Process	Day
A	Potassium oxide	Haber	Tuesday
B	Sodium hydroxide	Bayer	Monday
C	Magnesium	Kroll	Thursday
D	Vanadium(V) oxide	Contact	Friday
E	Calcium carbonate	Solvay	Wednesday

Extension

During trialling some institutions extended this work by asking students to find out more about the chemical processes.

Acknowledgement

This activity was suggested by Jacqui Clee.

47. Making ice

Time

This is best suited to an extended study.

Level

12 years and upward.

Curriculum links

Melting points and boiling points.

Group size

1–3.

Materials and equipment

Materials per group

- deionised water.

Equipment per group

- 100 and 250 cm^3 beakers
- insulating material
- –5 to +100°C thermometers
- access to a refrigerator and freezer
- safety glasses.

Safety

Eye protection must be worn.

Risk assessment

A risk assessment must be carried out for this activity.

Commentary

Students will find that hot water freezes more quickly than cold water – more precisely, water freezes more slowly if the initial temperature is below room temperature. The explanation is not entirely clear, but may be because a hot liquid has a 'hot top' of mobile molecules with high kinetic energy. These molecules can escape from the liquid phase more easily than colder molecules with lower kinetic energy in a cooler liquid. The rapid cooling of the hot liquid is due to the evaporation from this 'hot top'.

This activity is based on an article that appeared in *Physics Education*. Erasto Mpemba was a student at Magamba Secondary School in Tanzania and he discovered the phenomenon while making ice cream. One day in order to be sure of space in the refrigerator Mpemba put his ice cream mixture into the fridge without letting it cool first. At the same time one of his friends, who had let his mixture cool, also put his mixture into the fridge. To everyone's surprise Mpemba's ice cream froze first after about 1 h, while his friend's remained liquid for longer.

Francis Bacon[1] reported in 1620 that 'Water slightly warm is more easily frozen than quite cold', and some people may have come across the folklore 'never pour hot water down a frozen drain because the water will only freeze faster'.

The problem may challenge the perception of scientific 'facts'.

Reference

1. F. Bacon, *Novum Organian,* 1620.

Acknowledgement

This idea is based upon an article by Martin Sherwood.

48. Fizzy drinks

Time

1 h.

Level

12 years and upward.

Curriculum links

Hypothesis testing.

Group size

2– 4.

Materials and equipment

Materials per group

- cans of fizzy drink
- anti-bumping granules
- sand or other particulate material.

Equipment per group

- –5 to +100°C thermometer
- access to water baths and a refrigerator
- safety glasses.

Safety

Eye protection must be worn.

Risk assessment

A risk assessment must be carried out for this activity.

Commentary

This phenomenon is widely known and occurs with any substance that is under pressure. There are a number of variables to control and monitor: temperature, degree of shaking, size of can *etc*, and it is possible to build up a set of conditions that are known to maximise the degree of frothing and to minimise it.

The explanation for this phenomenon is based on nucleation. It is difficult to determine this from experimentation. When the pressurised can is shaken, small vapour bubbles of gas enter the drink and these act as a 'nucleus' for dissolved gas. When the can is opened the pressure drops and the bubbles expand rapidly, shooting to the surface causing the drink to froth out of the can. The critical step for students is to sprinkle sand (or similar particulate material) into the fizzy drink – bubbles immediately form around these small particles and hence give a clue to the formation of bubbles when the can is shaken.

Students may be interested to link this phenomenon with (a) the principle of crystallisation and (b) how anti-bumping granules work (or glass coasters in milk saucepans).

Acknowledgement

This activity is based on a suggestion from Jim Iley.

Esso

49. After meal puzzle

Time

15 min.

Level

Suitable for most ages.

Curriculum links

A knowledge of freezing point depression in solutions is helpful but not essential.

Group size

1–2.

Materials and equipment

Materials per group

- ice cubes
- water
- common salt (NaCl).

Equipment per group

- glass tumbler
- plastic cup or beaker
- 10 cm of string.

Safety

There are no particular safety issues.

Commentary

This simple puzzle is based on an experiment in a book called *After dinner science.*[1] It might be a memorable way of introducing the depression of freezing point.

Procedure

The problem can be solved by first soaking the end of the string in the water. The end is then laid across the top of the ice cube and a little salt is sprinkled along each side of the string. The salt lowers the freezing point of the ice that it touches, causing it to melt. The heat needed to melt the ice is withdrawn from the adjacent ice and from the water on the string. Within seconds the string should become frozen to the ice cube so that it can be lifted up.

Extension

Do other soluble substances – *eg* sugar and 'low salt' – work as well as common salt?

Reference

1. K. Swezey, *After dinner science.* London: Nicholas Kaye, 1949.

50. A chemically powered boat: a bubble boat race

Time

The problem could be set 2 weeks or more before the day of testing.

One or two 2 h sessions will be needed for making and testing models. It is hoped that some students may become motivated to try out and redesign their models in their own time. They will need access to water to do this.

30 minutes for judging (if a competition).

Level

Accessible to a wide range of abilities.

Curriculum links

Designing and making skills, production of carbon dioxide gas, properties of gases.

Group size

3– 4.

Materials and equipment

Materials per group

- 3 level teaspoons of sodium hydrogencarbonate
- 9 level teaspoons of citric acid .

Equipment per group

- items from the junk list (pXX)
- identical small plastic teaspoons
- balloons
- balsa wood
- tools for working balsa wood
- woodwork (PVA) or modelling glue
- thin card (cereal packets, postcards, *etc*)
- plastic drinking straws (straight and bendy)
- paper clips
- cocktail sticks
- plastic tubing (tubes from old ball-point pens might be useful)
- expanded polystyrene tiles
- sewing thread
- safety glasses
- access to an area of water.

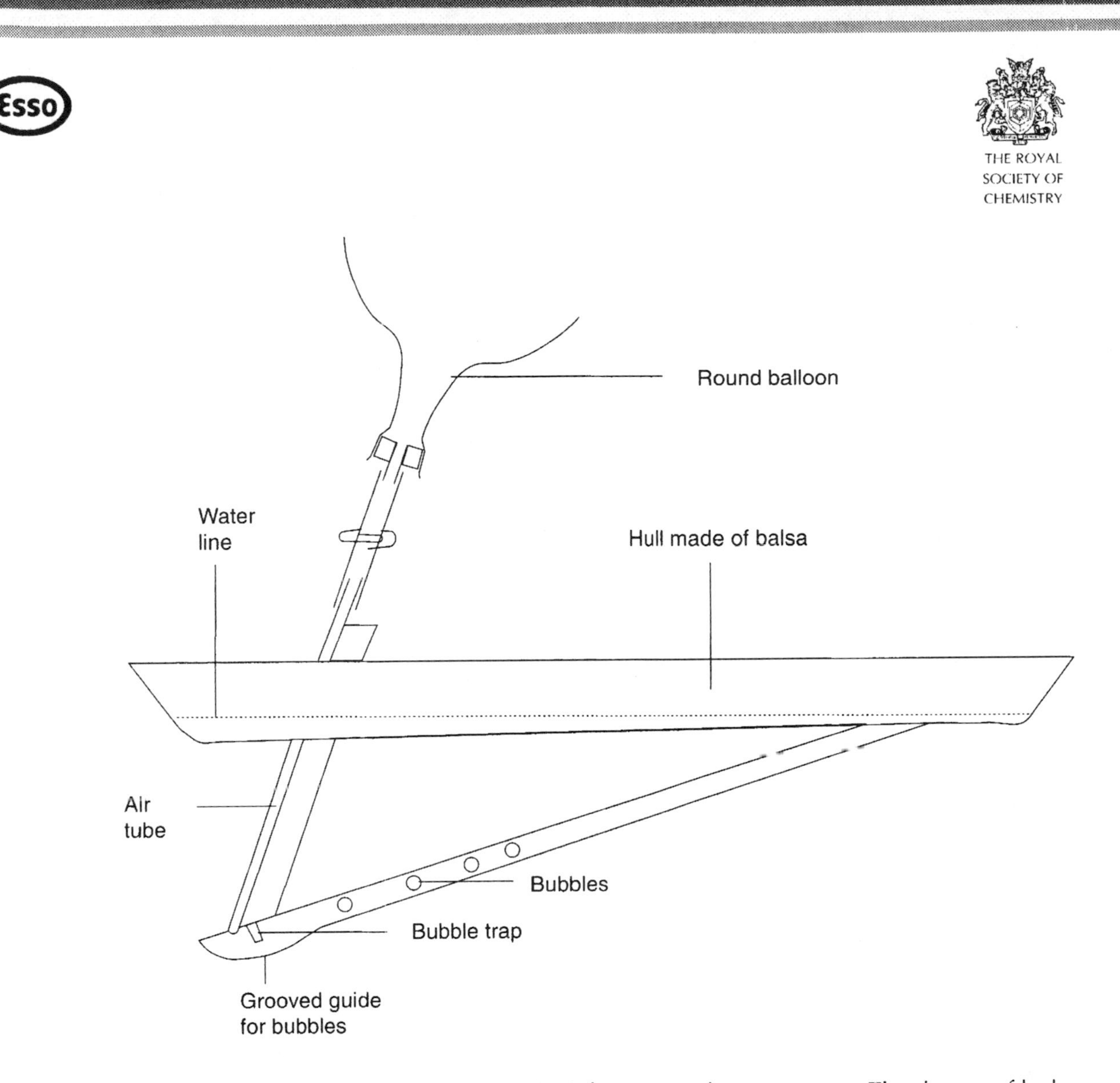

Boats using the 'bubble power' principle can travel many metres. The danger of leaks and floods is likely to be a problem on upper floors. Sections of rain guttering of 3–4 m in length make a good race course but the guttering must be chosen with a deep cross-section if it is to take the deep-keeled bubble boat in the photograph. For preliminary trials a large sink may be used. Very large photographic dishes are also suitable.

Safety

Eye protection must be worn.

Risk assessment

A risk assessment must be carried out for this activity.

Commentary

This egg race is an extension of the boat race described in a previous publication.[1] The designs reported so far have all been based on jet propulsion.[2,3] The boats (often floating plastic bottles) zoom along the surface of the water, even tending to take off; however, the gas is soon exhausted and they do not go very far. Models built on a similar principle will travel along a flat solid surface.

This challenge uses 'bubble power' to drive the model. The version in the photograph is a catamaran which can travel at a stately pace for over 30 minutes and cover a considerable distance. It was one of several models built by Alan Stevens of Loughborough University. The design is based on the Bubble Cat catamaran[4] in which the bubbles of gas slide backwards up a wedge-shaped keel. The buoyancy of

THE ROYAL
SOCIETY OF
CHEMISTRY

the bubbles produces a forward thrust on the keel and the boat is designed to reduce the drag of the water as much as possible. During trialling students came up with versions of their own and in one institution the problem became a joint venture between the science and technology departments.

Procedure

This activity is designed to be fun. The drawings and photograph are included to suggest a possible approach. Complete instructions for a successful model are given in the book referred to above.[4]

Extension

The students may check the stoichiometry of the reactants and may try varying the quantities to inflate a balloon an optimum amount. They may also like to adapt the model so that it can be loaded with chemicals at the start.

References

1. K. Davies, *In search of solutions*. London: RSC, 1990.
2. L. J. Wygoda, *Chem. 13 News*, September 1992.
3. W. Martin, *The Science Teacher*, May 37, 1991.
4. P. Holland, *Amazing models! – balloon power*. London: Argus Books, 1989.

Acknowledgements

Alan Stevens of Loughborough University produced the first successful model boat. Berinda Banks and her pupils at Mill Hill School designed and built other versions.

THE ROYAL
SOCIETY OF
CHEMISTRY

Printed in the United Kingdom
by Lightning Source UK Ltd.
115368UKS00001B/17-18